Studies in Computational Intelligence

Volume 795

Series editor

Janusz Kacprzyk, Polish Academy of Sciences, Warsaw, Poland
e-mail: kacprzyk@ibspan.waw.pl

The series "Studies in Computational Intelligence" (SCI) publishes new developments and advances in the various areas of computational intelligence—quickly and with a high quality. The intent is to cover the theory, applications, and design methods of computational intelligence, as embedded in the fields of engineering, computer science, physics and life sciences, as well as the methodologies behind them. The series contains monographs, lecture notes and edited volumes in computational intelligence spanning the areas of neural networks, connectionist systems, genetic algorithms, evolutionary computation, artificial intelligence, cellular automata, self-organizing systems, soft computing, fuzzy systems, and hybrid intelligent systems. Of particular value to both the contributors and the readership are the short publication timeframe and the world-wide distribution, which enable both wide and rapid dissemination of research output.

More information about this series at http://www.springer.com/series/7092

Stefka Fidanova

Editor

Recent Advances in Computational Optimization

Results of the Workshop on Computational Optimization WCO 2017

 Springer

Editor
Stefka Fidanova
Department of Parallel Algorithms
Institute of Information and Communication
  Technologies, Bulgarian Academy
  of Sciences
Sofia, Bulgaria

ISSN 1860-949X   ISSN 1860-9503   (electronic)
Studies in Computational Intelligence
ISBN 978-3-030-07618-4   ISBN 978-3-319-99648-6   (eBook)
https://doi.org/10.1007/978-3-319-99648-6

This Springer imprint is published by the registered company Springer Nature Switzerland AG
The registered company address is: Gewerbestrasse 11, 6330 Cham, Switzerland

Organization

Workshop on Computational Optimization (WCO 2017) is organized in the framework of Federated Conference on Computer Science and Information Systems (FedCSIS—2017).

Conference Co-chairs

Stefka Fidanova, IICT (Bulgarian Academy of Sciences, Bulgaria)
Antonio Mucherino, IRISA (Rennes, France)
Daniela Zaharie, West University of Timisoara (Romania)

Program Committee

Tibérius Bonates, Universidade Federal do Ceará, Brazil
Mihaela Breaban, University of Iasi, Romania
Camelia Chira, Technical University of Cluj-Napoca, Romania
Douglas Gonçalves, Universidade Federal de Santa Catarina, Brazil
Hiroshi Hosobe, National Institute of Informatics, Japan
Hideaki Iiduka, Kyushu Institute of Technology, Japan
Carlile Lavor, IMECC-UNICAMP, Campinas, Brazil
Pencho Marinov, Bulgarian Academy of Sciences, Bulgaria
Flavia Micota, West University Timisoara, Romania
Ionel Muscalagiu, Politehnica University Timisoara, Romania
Konstantinos Parsopoulos, University of Patras, Greece
Camelia Pintea, Technical University Cluj-Napoca, Romania
Olympia Roeva, Institute of Biophysics and Biomedical Engineering, Bulgaria
Patrick Siarry, Universite Paris XII Val de Marne, France
Stefan Stefanov, Neofit Rilski University, Bulgaria

Ruxandra Stoean, University of Craiova, Romania
Catalin Stoean, University of Craiova, Romania
Tomas Stuetzle, Universite libre de Bruxelles, Belgium
Tami Tamir, the Interdisciplinary Center (IDC), Israel
Antanas Zilinskas, Vilnius University, Lithuania

Preface

Many real-world problems arising in engineering, economics, medicine, and other domains can be formulated as optimization tasks. Every day, we solve optimization problems. Optimization occurs in the minimizing time and cost or the maximization of the profit, quality, and efficiency. Such problems are frequently characterized by non-convex, non-differentiable, discontinuous, noisy, or dynamic objective functions and constraints which ask for adequate computational methods.

This volume is a result of very vivid and fruitful discussions held during the Workshop on Computational Optimization. The participants have agreed that the relevance of the conference topic and quality of the contributions have clearly suggested that a more comprehensive collection of extended contributions devoted to the area would be very welcome and would certainly contribute to a wider exposure and proliferation of the field and ideas.

The volume includes important real problems like modeling of physical processes, wildfire modeling, modeling processes in chemical engineering, workforce planning, wireless access network topology, parameter settings for controlling different processes, expert knowledge, berth allocation, identification of homogeneous domains, quantum computing. Some of them can be solved by applying traditional numerical methods, but others needs a huge amount of computational resources. Therefore for them are more appropriate to develop algorithms based on some metaheuristic method like evolutionary computation, ant colony optimization, particle swarm optimization, constraint programming.

Sofia, Bulgaria
April 2018

Stefka Fidanova
Co-Chair, WCO'2017

Contents

Contraction Methods for Correlation Clustering: The Order is Important

László Aszalós and Mária Bakó

Abstract Correlation clustering is a NP-hard problem, and for large graphs finding even just a good approximation of the optimal solution is a hard task. In previous articles we have suggested a contraction method and its divide and conquer variant. In this article we examine the effect of executing the steps of the contraction method in a different order.

1 Introduction

Clustering is an important tool of unsupervised learning. Its task is to group objects in such a way, that the objects in one group (cluster) are similar, and the objects from different groups are dissimilar. It generates an equivalence relation: the objects being in the same cluster. The similarity of objects are mostly determined by their distances, and the clustering methods are based on distance.

Correlation clustering is an exception, it uses a tolerance (reflexive and symmetric) relation. Moreover it assigns a cost to each partition (equivalence relation), i.e. number of pairs of similar objects that are in different clusters plus number of pairs of dissimilar objects that are in the same cluster. Our task is to find the partition with the minimal cost. Zahn proposed this problem in 1965, but using a very different approach [13]. The main question was the following: *which equivalence relation is the closest to a given tolerance relation?* Many years later Bansal et al. published a paper, proving several of its properties, and gave a fast, but not quite optimal algorithm to solve this problem [6]. Bansal have shown, that this is an NP-hard problem.

L. Aszalós (✉)
Faculty of Informatics, University of Debrecen, 26 Kassai str.,
Debrecen H4028, Hungary
e-mail: aszalos.laszlo@inf.unideb.hu

M. Bakó
Faculty of Economics at University of Debrecen, 138 Böszörményi str.,
Debrecen H4032, Hungary
e-mail: bakom@unideb.hu

© Springer Nature Switzerland AG 2019
S. Fidanova (ed.), *Recent Advances in Computational Optimization*,
Studies in Computational Intelligence 795,
https://doi.org/10.1007/978-3-319-99648-6_1

1

The number of equivalence relations of n objects, i.e. the number of partitions of a set containing n elements is given by Bell numbers B_n, where $B_1 = 1$ and $B_n = \sum_{i=1}^{n-1} \binom{n-1}{k} B_k$. It can be easily checked that the Bell numbers grow exponentially. Therefore if $n > 15$, in a general case we cannot achieve the optimal partition by exhaustive search. Thus we need to use some optimization methods, which do not give optimal solutions, but help us achieve a near-optimal one.

If the correlation clustering is expressed as an optimization problem, the traditional optimization methods (hill-climbing, genetic algorithm, simulated annealing, etc.) could be used in order to solve it. We have implemented and compared the results in [1].

This kind of clustering has many applications: image segmentation [10], identification of biologically relevant groups of genes [7], examination of social coalitions [12], improvement of recommendation systems [9] reduction of energy consumption [8], modelling physical processes [11], (soft) classification [3, 4], etc.

In a former article we have shown the clustering algorithm based on the divide and conquer method, which was more effective than our previous methods. But our measurements have pointed out, that this method is not scalable. Hence for large graphs the method will be very slow. Therefore we would like to speed up the method. The *simplest* way to do this would be to distribute the calculations between the cores of a processor. Unfortunately, theory and practice often differs.

The structure of the paper is the following: in Sect. 2 correlation clustering is defined mathematically, Sect. 3 presents the contraction method and some variants. Next, the best combination of local improvements is selected, and in Sect. 5 the former divide and conquer method is improved. Later the technical details of the concurrency is discussed.

2 Correlation Clustering

In the paper the following notations are used: V denotes the set of the objects, and $T \subset V \times V$ the tolerance relation defined on V. A partition is handled as a function $p : V \rightarrow \{1, \ldots, n\}$.

The objects x and y are in a common cluster, if $p(x) = p(y)$. We say that objects x and y are in conflict at given tolerance relation and partition iff value of $c_T^p(x, y) = 1$ in (1).

$$c_T^p(x, y) \leftarrow \begin{cases} 1 \text{ if } (x, y) \in T \text{ and } p(x) \neq p(y) \\ 1 \text{ if } (x, y) \notin T \text{ and } p(x) = p(y) \\ 0 \text{ otherwise} \end{cases} \tag{1}$$

We define the cost function of relation T according to partition p as:

$$c_T(p) \leftarrow \frac{1}{2} \sum c_T^p(x, y) = \sum_{x<y} c_T^p(x, y) \tag{2}$$

The task of correlation clustering is to determine the value of $\min_p c_T(p)$, and a partition p for which $c_T(p)$ is minimal. Unfortunately, this exact value cannot be determined in practical cases, except for some very special tolerance relations. Hence we can only get approximative, near optimal solutions.

Correlation clustering can be defined as a problem of statistical physics [11], where the authors use analogies from physics to solve the problem for small graphs. Here we do something similar. We can define the *attraction* between two objects: if they are similar then the attraction between them is 1; if they are dissimilar then the attraction between them is -1 (they repulse each other); otherwise—which can occurs at a partial tolerance relation—the attraction is 0.

$$a(x, y) \leftarrow \begin{cases} 1, \text{ if } (x, y) \in T \\ -1, \text{ if } (x, y) \notin T \\ 0, \text{ otherwise} \end{cases} \tag{3}$$

(3) can be generalized for object x and for clusters g and h:

$$a'(x, g) = \sum_{y \in g} a(x, y) \text{ and } \hat{a}(g, h) = \sum_{y \in h} a'(y, g).$$

We leave it to the reader to check that if these sums are positive and we join this element and cluster, or these clusters—by getting a partition p' containing the clusters $g \cup \{x\}$ or $g \cup h$—then $c_T(p) \geq c_T(p')$. This means that by joining attractive clusters, the cost decreases.

The left part of Fig. 1 shows a tolerance relation. Here the black pixels denote the similarity and the light grey pixels denote the dissimilarity. This tolerance relation belongs to a Barabási-Albert type random graph (BA), where most pairs of objects are unrelated (denoted by white pixels). The right part of this figure present a near

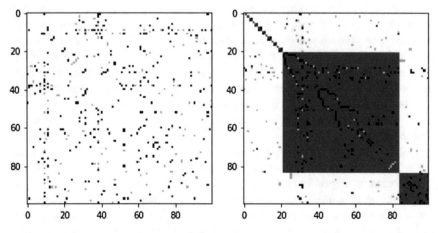

Fig. 1 A correlation clustering problem (on left) and its near optimal solution (on right)

optimal solution for this problem: we reordered the rows and columns of the previous picture, and marked the clusters with grey. Hence the conflicts are the following: black pixels outside the grey squares (similar objects in different clusters) and light grey pixels in the grey squares (dissimilar objects in common clusters).

3 Contraction Method

The contraction method [5] is based on two operation: contraction and movement. The name *contraction* means that we join two attractive clusters. We can treat a cluster as *stable*, if for each of its elements the best position is inside this cluster, because the superposition of the forces—attraction or repulsion between the given element and other elements—is attraction for each element in the cluster; and there does not exist another cluster which is more attractive for any element in the cluster than its own cluster. It is possible that the joining of two stable clusters produces a non-stable cluster: the new elements are mostly repulsive for some element. In this case to get less conflicts this element needs to be *moved* into another cluster. Specially, if one object is repulsed by all clusters, a singleton containing this element needs to be constructed. The process includes calculation of the attractive forces of one object x for all clusters, and moving nodes based on the maximal attraction we called *movement*.

Aszalós and Mihálydeák [5] contains the forces that need to be recalculated after a movement or a contraction. These recalculations can be applied for any kind of tolerance relation. If the graph of the tolerance relation is dense—by using the matrices of forces between objects, forces between objects and clusters (for the movements) and forces between clusters (for the contractions)—then the contraction method can be run in an efficient way by adding and subtracting rows and columns of these matrices.

If the graph of the tolerance relation is sparse, then it is a waste to use full matrices for storing the actual forces. (If the tolerance relation contains small amounts of dissimilarity, then the optimal partition consist of only some clusters, so a small matrix is enough to store the forces between clusters.) In our former articles the algorithms were implemented in Python, and we used associative arrays (dict) and associative arrays of associative arrays to store the non-zero objects. But the deletion is problematic for this data type, because the implementations based on hash only apply logical deletion. Working with big tolerance relations, the limit of the implementation is noticeable.

Our new implementation approaches the problem from a new direction. By working with a sparse graph, most of its nodes (the objects) only have a few neighbours. If the forces on a specific node need to be calculated (for the movement), only its neighbours need to be checked, not all of the objects. Therefore, instead of searching for the neighbours of a given object again and again, we store them and the signs of their edges. Of course this means each edge is stored twice i.e. at both of its endpoints, but at a sparse graph—where $|E| = O(|V|)$—this is not a serious problem.

To calculate the forces between clusters in case of a dense graph all edges need to be visited, so the complexity is $O(n^2)$. But at sparse graphs the number of edges is proportional to the number of vertices, so the complexity of the calculation of forces is $O(n)$.

To economize the calculation we can examine what forces will change after a contraction or a movement. After a movement the given object will be moved into another cluster, and this only effects its neighbours. At BA graphs' degree distribution follows the power law, but the average of the degrees is constant. Therefore the amount of force recalculation is near constant. At contraction two clusters are joined, and we need to recalculate the forces for every neighbour of the objects that are now in the joined cluster. Depending on the size of this new cluster this can cause few to many recalculations. It is worth to examine the number of recalculations and if there are many it is economical to execute a total recalculation.

Correlation clustering can be treated as an optimization problem, where the aim is to minimize the number of conflicts. The steps of contraction and movement can be treated as a local search step. Nevertheless, simple variants of the hill climbing method are not effective for this problem. We have tested this in case of a graph with 13 nodes and from almost 3 million partitions only 2 were global optimum and around ten thousand were local minimum. For bigger graphs the ratio of numbers of global and local optima will be even smaller, so to find a global optima or a near optimal local optima is a truly difficult task.

The interesting question is *how to combine the steps of contraction and movement?* Figure 2a shows the algorithm we implemented in the former article [2]. A contraction could be a dramatic change, like when two big clusters are joined. This means that with contraction many object get their final positions at the same time. In this

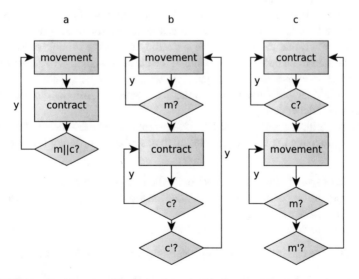

Fig. 2 Different algorithms (combinations of steps) of the local search

variant this contraction step is repeated until it is successful (the number of conflicts decreases). Next, from the unstable clusters some objects are moved to better positions, and this cycle is repeated until it is profitable. This algorithm produced a rather fast method. By rewriting the source we compared this algorithm with some other variants combining the steps of contraction and movement.

At first we tested what is the effect of changing the order of contraction and movement. Figure 2b shows a variant with a different order, where we execute a contraction after each movement. It is obvious, that the movements only produce local changes, so it takes many cycles to move the objects into their final cluster.

Finally we created a variant which moves the objects until it is profitable, then joins the clusters until is it profitable, and if there was a contraction, then it starts a new cycle, as Fig. 2c shows. It is obvious that this variant is the fastest: the contraction goes in big steps towards the optimum, it can reduce the number of cycles, hence speeding up the calculations. But what is the price of this?

Our experiments show that there is an unbreakable conflict between the speed and efficiency/accuracy: the number of conflicts at method c were 13% less that at method a.

4 Quasi-parallel Variant

Formerly we have discussed the quasi-parallel variant of the algorithms [5]. The most naive variant of the *contraction* step calculates all the forces between clusters, and next joins the two most attractive clusters and drops the other calculations. A bit cleverer variant reuses the calculated forces to calculate the forces according to the new (contracted) clusters.

The most costive variant wants to use all the calculations (without any recalculations). Hence it sorts all the calculated forces in decreasing order—as it is a greedy algorithm—and if that value is positive and valid, it joins the suitable clusters. When can a calculated force be invalid? If some of the clusters it belongs to do not exist any more, because we have already merged them with another, a third cluster. We named this last variant as *quasi-parallel*, because we practically join the clusters parallel, although not independently.

On https://github.com/aszalosl/DC-CC2 we published the sources of our experiments. We are using the following notation to describe the different kind of contractions:

O: we use only one, the best contraction (called `one` in the sources)
R: we apply the best contraction, recalculate the forces and repeat until it is profitable (called `iterate_min`)
I: the quasi-parallel version (called `independent`)

Of course we can implement similar variants for the *movement* step too:

a: we move all objects—sequentially—into its optimal cluster, i.e. we recalculate the forces for each object (called `the_all`)

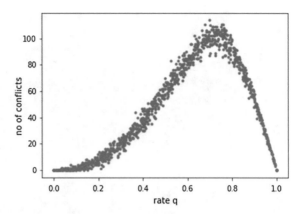

Fig. 3 Number of conflicts at different rates q

i: the quasi-parallel movement—without recalculation—(called `independent`)

Finally Fig. 2 describes three algorithms:

a: called `sequence`
b: called `iterated`
c: called `iterated2`

Figure 3 presents the number of conflicts of a thousand correlation cluster problems. Here the axis x denotes the rate of positive edges among all edges. From the figure it is obvious that this number does not determine the graph, the distribution of these positive edges have a big impact. At dense Erdős-Rényi graphs (ER) this area is thicker, because any pairs of nodes are connected with probability p (uniformly), hence this generation results in more symmetric graphs.

If q is too small (most of the edges are repulsive), then in the solution almost all clusters became small, and the preprocessing had no profit. If q is close to 1.0 (most of the edges are attractive), then we only have a few clusters in the solution, and one of them contains almost all the nodes. In this case the preprocessing is very profitable.

The band in Fig. 3 is at $q \approx 0.71$ is the widest, and the number of conflict is the highest. This means that at this type of (random) graphs this value is the most problematic one.

Although we could create a $3 \times 2 \times 3$ combination from the algorithm, the contraction and the movement, the first variant of the contraction (O) is not so important—even if it is in a cycle, which is implemented inside R—hence we show in Fig. 4 only the $2 \times 2 \times 3 + 1$ variant. From top to bottom the combinations are the following: *baI, baR, biI, biR, caI, caR, ciI, ciR, aaI, aaR, aiI, aiR*, and finally *aaO*. We have used 100 different values of q from 0.0 to 1.0, to generate Fig. 4. In each column the best value is denoted with white and the worst with black. The shade of grey is proportional to the goodness of the generated results. We made several experiments and used their average, to exclude any accidental results.

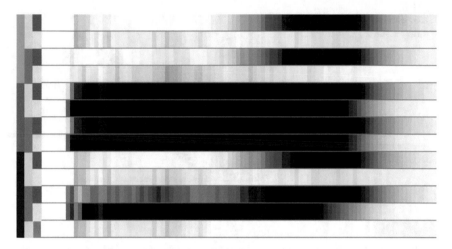

Fig. 4 Effectiveness of the different combinations of movement and contraction

At the end of the previous section we have mentioned that the algorithm c gives poor results, as shown in Fig. 4. By examining the results of algorithms a and b we can realize, that the quasi-parallel movement gives worse result than the movement with total calculation, mainly around the most problematic value of q. To compare algorithms a and b using movement with total calculation, we need a more accurate method.

Hence we measured the time needed to execute the different methods and we summed the conflicts at different values of q. The results are presented in Fig. 5: on the left the numbers, and on the right the same data visualized. We ordered the methods by their efficiency.

From the methods *aiO, biR, aaR* and *baI* are the Pareto optimum. We can choose from them by the expected accuracy and efficiency. The quasi-parallel calculations could help with the speed and accuracy, but we want more.

5   Divide and Conquer Reloaded

In a former article we have examined whether the divide and conquer approach is useful for correlation clustering [2]. As a reminder, the divide and conquer solution consist of three *simple* steps:

- divide the problem into subproblems,
- solve the subproblems (in a recursive way)
- construct the solution of the original problem from the solutions of the subproblems.

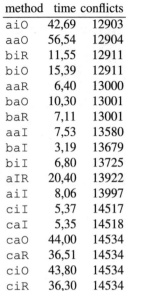

method	time	conflicts
aiO	42,69	12903
aaO	56,54	12904
biR	11,55	12911
biO	15,39	12911
aaR	6,40	13000
baO	10,30	13001
baR	7,11	13001
aaI	7,53	13580
baI	3,19	13679
biI	6,80	13725
aIR	20,40	13922
aiI	8,06	13997
ciI	5,37	14517
caI	5,35	14518
caO	44,00	14534
caR	36,51	14534
ciO	43,80	14534
ciR	36,30	14534

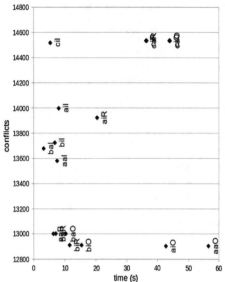

Fig. 5 Speed versus efficiency

In some cases some of the steps could be left out or are very trivial. In our former article the construction of the sub-problems was simple: we divided the graph into same size sub-graphs by the IDs of the objects. It can be checked easily, that with this construction most of the edges are left out from our calculations.

In case of ER random graphs the edges are distributed uniformly at random. As the matrices of subgraphs cover only n/n^2 part of the matrix of original graph, only $1/n$ of the edges are left to work with. We construct slightly better sub-graphs with a little effort i.e. with complexity of $O(n)$. The breadth first traversal of the graph is used, and the nodes are taken in that order in which they are deleted from the fringe, i.e. when they get closed. The effectiveness of this trick is shown in Fig. 6. On the left we give the numbers and on the right the grey part denotes the remaining edges. It is not a surprise that in case of ER graphs, where the edges are independent from each other, this trick has no real effect. But at BA random graphs—where the construction guaranties that the edges are not independent—the trick works well: when the former method left 5% of the edges, this trick leaves us 40% of them.

The other steps of the divide and conquer approach remained the same. The sub-problems were solved by recursion if they were big enough, otherwise a direct solution was used: starting from singleton clusters, the algorithm of Fig. 2c for the graph of the sub-problem was followed. Finally all the clusters from the solutions of the sub-problems were collected and put together (as an initial clustering of the whole graph), and we executed the algorithm of Fig. 2c again. It is surprising, but *solving the original problem, the sub-problems, the subsub-problems, etc. is faster*

N	no. of subgraphs			
	2	5	10	20
	BA(3,2)			
500	62.8/50.6	45.1/18.9	43.4/9.8	39.1/5.7
1000	63.2/48.9	50.3/21.6	42.8/9.8	40.4/5.2
5000	62.5/50.0	49.3/20.9	45.3/10.0	42.6/4.5
10000	62.6/50.3	49.2/20.0	44.8/10.6	42.8/5.0
20000	62.7/50.2	49.5/20.4	44.7/10.1	42.8/4.9
	ER (p=0.015)			
500	58.6/51.2	39.0/19.7	31.1/9.4	28.2/5.5
1000	53.4/49.9	28.1/20.6	20.8/10.2	16.2/4.8
5000	51.2/50.0	21.7/20.0	11.9/10.0	7.3/4.9
10000	50.5/49.9	20.9/19.9	11.0/10.0	6.2/5.0
20000	50.2/50.0	20.4/20.0	10.5/10.0	5.6/5.0

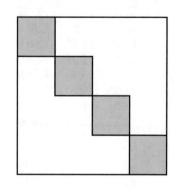

Fig. 6 Remaining edges of the subgraphs (in %)

than solving the original problem alone. This is not a paradox, the key question is the initial clustering of the original problem.

6 Technical Details: Concurrency

The last sentences of the previous chapter are very promising. Moreover at the reimplementation of our software we have taken care of parallelism.

We implemented our software in Python. For calculation intensive tasks this language offers a multiprocessing package. If the processes need to wait for each other, then we will not gain anything. Our processes need to work independently, hence we modified our software, and now we calculate the forces for moving in advance.

At first we used the instruction map for each object, which could be familiar to the reader from Google's MapReduce concept. We recall, that *movement* in the algorithm of Fig. 7 is inside a double loop (at Base case). One task (to calculate the forces on an object) is extremely simple, hence the overhead is huge, it runs thousands times slower than the original. Next we created a pool, and the set of nodes were divided into three or four (case Par3), and each core of the processor received one subset, and the role to calculate the forces on nodes that are in that subset.

Our framework—constructed for divide and conquer method (DC)—enables us to break the original problem into sub-problems, and solve them in parallel using the possibilities of a multi-core processor. Unfortunately each level of recursion adds some error to the calculations, hence the recursion is not precisely accurate. Instead we need to divide the original set into many smaller sets, and execute the method on each of them and in a next step execute the algorithm on the whole population. We

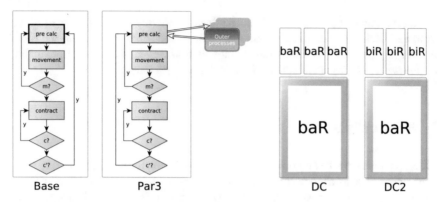

Fig. 7 The original method and the parallel versions

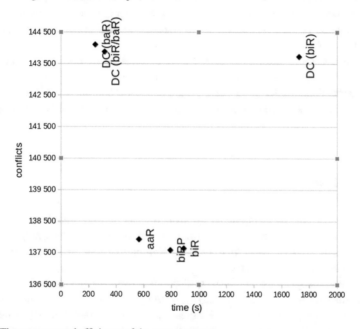

Fig. 8 The accuracy and efficiency of the concurrency

have the option to use same (DC) or different methods (DC2) on the lower and on the upper level.

Figure 8 presents the results of our experiments. The closeness of $biRP$ (Par3) and biR (Base) shows that at this method the concurrency does not give many advantages. Here the DC variants are far enough from the Base. This means, that the divide and conquer method collects some error from the partial solutions, but the running time decreases. Unfortunately when we have many objects, the distributed solving of the lower level is done almost immediately, but to combine the sub-solutions by solving the upper level problem takes a long time.

At problematic cases ($q = 0.71$) the hardness of solving the sub-problems brings overhead back. We examined the running time of the subproblems, and we found that for big graphs the combination of sub-solutions (repeat the contraction method for the whole graph) could take up 98% of the running time.

7 Future Plans

Although we have a fast algorithm to solve the problems for large graphs, and some hints about how to choose between them, the research is not over. When solving big graph problems, most of the time only one thread is running, hence we have possibilities to use the concurrency. It is worth to try a manager and a pool of worker processes defined not inside cycles, but at the upper levels. The overhead of the communication between processes could be problematic, but only tests could decide on usefulness of this approach.

The fastest computation is *no* computation. Therefore we need to examine which calculations are necessary, and which can be omitted.

Of course these tricks do not change the quadratic complexity of the algorithm, but we believe, that we can reduce the the constant part, which will be very important in practice.

8 Conclusion

We introduced a correlation clustering problem, and we presented the contraction method to solve it. We improved our former algorithm in several ways, and we created several variants to it. Some of them used the elements of concurrent execution of the Python code with a small success.

To our knowledge, these are the state of the art algorithms in correlation clustering.

We made several measurements and the results gave hints on how to select amongst them to solve a particular problems. By these measurements our method has quadratic complexity. Finally, we presented the bottleneck of the algorithms. Our next step is to eliminate this, hopefully by using concurrency in a different way.

References

1. Aszalós, L., Bakó, M.: Advanced search methods (in Hungarian). http://www.tankonyvtar.hu/hu/tartalom/tamop412A/2011-0103_13_fejlett_keresoalgoritmusok (2012)
2. Aszalós, L., Bakó, M.: Correlation clustering: divide and conquer. In: Ganzha, M., Maciaszek, L., Paprzycki, M. (eds.) Position Papers of the 2016 Federated Conference on Computer Science and Information Systems, *Annals of Computer Science and Information Systems*, vol. 9, pp. 73–78. PTI (2016). https://doi.org/10.15439/2016F168

3. Aszalós, L., Mihálydeák, T.: Rough clustering generated by correlation clustering. In: Rough Sets, Fuzzy Sets, Data Mining, and Granular Computing, pp. 315–324. Springer, Berlin, Heidelberg (2013). https://doi.org/10.1109/TKDE.2007.1061
4. Aszalós, L., Mihálydeák, T.: Rough classification based on correlation clustering. In: Rough Sets and Knowledge Technology, pp. 399–410. Springer (2014). https://doi.org/10.1007/978-3-319-11740-9_37
5. Aszalós, L., Mihálydeák, T.: Correlation clustering by contraction, a more effective method. In: Recent Advances in Computational Optimization, vol. 655, pp. 81–95. Springer (2016). https://doi.org/10.1007/978-3-319-40132-4_6
6. Bansal, N., Blum, A., Chawla, S.: Correlation clustering. Mach. Learn. **56**(1–3), 89–113 (2004). https://doi.org/10.1023/B:MACH.0000033116.57574.95
7. Bhattacharya, A., De, R.K.: Divisive correlation clustering algorithm (dcca) for grouping of genes: detecting varying patterns in expression profiles. Bioinformatics **24**(11), 1359–1366 (2008). https://doi.org/10.1093/bioinformatics/btn133
8. Chen, Z., Yang, S., Li, L., Xie, Z.: A clustering approximation mechanism based on data spatial correlation in wireless sensor networks. In: Wireless Telecommunications Symposium (WTS), 2010, pp. 1–7. IEEE (2010). https://doi.org/10.1109/WTS.2010.5479626
9. DuBois, T., Golbeck, J., Kleint, J., Srinivasan, A.: Improving recommendation accuracy by clustering social networks with trust. Recommender Systems & the Social Web **532**, 1–8 (2009). https://doi.org/10.1145/2661829.2662085
10. Kim, S., Nowozin, S., Kohli, P., Yoo, C.D.: Higher-order correlation clustering for image segmentation. In: Advances in Neural Information Processing Systems, pp. 1530–1538 (2011). DOI 10.1.1.229.4144
11. Néda, Z., Florian, R., Ravasz, M., Libál, A., Györgyi, G.: Phase transition in an optimal clusterization model. Physica A: Stat. Mech. Appl. **362**(2), 357–368 (2006). https://doi.org/10.1016/j.physa.2005.08.008
12. Yang, B., Cheung, W.K., Liu, J.: Community mining from signed social networks. Knowl. Data Eng. IEEE Trans. **19**(10), 1333–1348 (2007)
13. Zahn Jr, C.: Approximating symmetric relations by equivalence relations. J. Soc. Ind. Appl. Math. **12**(4), 840–847 (1964). https://doi.org/10.1137/0112071

Optimisation of Preparedness and Response of Health Services in Major Crises Using the IMPRESS Platform

Nina Dobrinkova, Thomas Finnie, James Thompson, Ian Hall,
Christos Dimopoulos, George Boustras, Yianna Danidou,
Nectarios Efstathiou, Chrysostomos Psaroudakis, Nikolaos Koutras,
George Eftichidis, Ilias Gkotsis, Marcel Heckel, Andrej Olunczek, Ralf Hedel,
Antonis Kostaridis, Marios Moutzouris, Simona Panunzi, Geert Seynaeve,
Sofia Tsekeridou and Danae Vergeti

N. Dobrinkova (✉)
IICT-BAS, Sofia, Bulgaria
e-mail: ninabox2002@gmail.com

T. Finnie · J. Thompson · I. Hall
Public Health England, London, UK
e-mail: thomas.finnie@phe.gov.uk

C. Dimopoulos · G. Boustras · Y. Danidou
European University of Cyprus -EUC, Nicosia, Cyprus
e-mail: C.Dimopoulos@euc.ac.cy

Y. Danidou
e-mail: Y.Danidou@external.euc.ac.cy

N. Efstathiou · C. Psaroudakis · N. Koutras
ADITESS, Nicosia, Cyprus
e-mail: nectarios@aditess.com

C. Psaroudakis
e-mail: cpsaroudakis@aditess.com

G. Eftichidis · I. Gkotsis
Center for Security Studies - KEMEA, Athens, Greece
e-mail: g.eftychidis@gmail.com

I. Gkotsis
e-mail: iliasgkotsis@gmail.com

M. Heckel · A. Olunczek · R. Hedel
IVI-Fraunhofer, Dresden, Germany
e-mail: marcel.heckel@ivi.fraunhofer.de

A. Olunczek
e-mail: andrej.olunczek@ivi.fraunhofer.de

A. Kostaridis · M. Moutzouris
SATWAYS, Athens, Greece
e-mail: a.kostaridis@satways.net

© Springer Nature Switzerland AG 2019
S. Fidanova (ed.), *Recent Advances in Computational Optimization*,
Studies in Computational Intelligence 795,
https://doi.org/10.1007/978-3-319-99648-6_2

15

Abstract We present the IMPRESS software system; a tool that can support the first responders in cases of disaster management and resources allocation and the optimisation of response to large-scale emergencies. This is a multi-level architecture system that has desktop and mobile interfaces, supporting the decision makers with different modules. Every module has been designed in a way that the data can be unified, no matter the source, and the related calculation engines are providing optimized information for both the users in the incident management centre and on the field. In this article IMPRESS system components are presented, among the validation and optimization activities during the demonstrations implemented in Palermo, Italy (field test exercise), Podgorica, Montenegro (field test exercise) and Sofia, Bulgaria (table top exercise).

1 Introduction

We introduce the IMPRESS system a command and control, incident management, and decision support system. This system has been developed to optimise the response of emergency workers in cases of extra-ordinary public health challenges (EOPHC). We distinguish EOPHC from the ordinary business of a health system as those circumstances that require a change from standing arrangements to those that require a qualitatively different response from the health services. Such circumstances are commonly referred to using a wide variety of ill-defined terms such as "major incident", "disaster", "crisis", etc. and may encompass events and incidents as varied as earthquakes, multiple vehicle collisions, chemical spills, terrorist activities, and so on.

The overall challenge to such a system is to minimise the number of people suffering life altering or terminal injury during a EOPHC. This requirement is set

M. Moutzouris
e-mail: m.moutzouris@satways.net

S. Panunzi
CNR, Rome, Italy
e-mail: simona.panunzi@biomatematica.it

G. Seynaeve
ECOMED, Brussels, Belgium
e-mail: geert.seynaeve@attentia.be

S. Tsekeridou · D. Vergeti
INTRASOFT International, Athens, Greece
e-mail: sofia.tsekeridou@intrasoft-intl.com

D. Vergeti
e-mail: danae.vergeti@intrasoft-intl.com

against a background of a resource constrained situation where it is unlikely to be possible to provide the same range of services and treatments that would be ordinarily available.

Both the situations that the IMPRESS system is envisaged to assist with, and the challenges within each domain that the system encompasses, are complex. Because of this the IMPRESS system is made up from a series of loosely coupled components each of which has been constructed to address a specific, domain specific challenge.

To reach an optimal solution each component must reach an optimal solution within its own domain and then for the system to reach a globally optimal solution based on the constraints imposed by the specific situation. In the following material we present each of the IMPRESS system components and the problems which they solve. These components, integrated into a single unified whole were then tested using a variety of live and table-top exercises. The outcomes from these exercises are systematically evaluated and results and conclusions drawn.

2   Core Components of the IMPRESS System

2.1   INCIMAG—Desk Based Command and Control Software and Primary IMPRESS System Frontend

The INCIMAG component is a fully fledged Integrated Command and Control Incident Management System. Its purpose is to interconnect stakeholders, decision makers, operators, first responders through standardized data interoperability for incident coordination and shared situational awareness. The primary focus is for Emergency Medical Services by implementing business logic and data specific to Emergency Medical Services and Hospital Emergency Department's such as incident representation, unit tasking, triage, bed availability, treatment and transport tracking of emergency patients. A typical deployment is shown in Fig. 1. The INCIMAG platform was built on the following key design principles:

- Daily use and during EOPHC situations for managing field and control center operations
- Support the three layers of command—The system can be used by all the various levels of command: strategic, tactical and operational
- Multi-Agency Coordination—to ensure that multiple agencies can coordinate during an EOPHC.

Upon the encounter of medical personnel with a victim, the victim is assigned unique identification that follows them from that point onward until their release from medical care. At each encounter their medical status is collected and a message is generated providing the current status of patient which help first responders, allow for control center dispatch personnel in decision making and provides an early-warning to hospitals in close proximity. Information kept on a patient is limited to what is required to provide adequate medical care.

One the main functionalities of the HIMS, according to the end user requirements, is the ability to monitor and communicate its available services and bed availability in real-time in order to assist a system/region's ability to care for a surge of patients in the event of an extraordinary public health emergency/mass casualty incident presented at Fig. 2.

INCIMAG provides for the interaction and integration of field mobile devices, in particular those running the INCIMOB software. The system provides: authentication of field devices against particular INCIMAG instances, situational awareness via visually and text feedback of field unit location, a mechanism for requests about patient status, tasking and dispatch for field units, and messaging an communication functions.

It is expected that in mass casualty incidents there is likely to be interactions between various agencies. This interaction between such agencies takes place through the Emergency Message Content Router (EMCR). This software route messages to the proper recipients who have declared their interest in particular messages, roles, keywords or geographic areas. Recipients subscribe and connect with authentication ensuring only known recipients receive particular messages. The operators can select directly which INCIMAG to send particular information based on their subscription criteria. Alternatively, the message is broadcasted and INCIMAG's that match the message's criteria vs their subscription criteria will receive the message an example of EMCR usage is shown on Fig. 3.

After patients have been registered within the system, either using INCIMOB or directly, an INCIMAG operator can trigger the Recommendation Engine to calculate an optimal distribution of patients to hospitals. In turn, the results will be sent and displayed to the related INCIMOB. With the help of this information, the INCIMOB unit knows which patient is to be transported to which hospital.

Fig. 1 A typical dual-monitor INCIMAG configuration

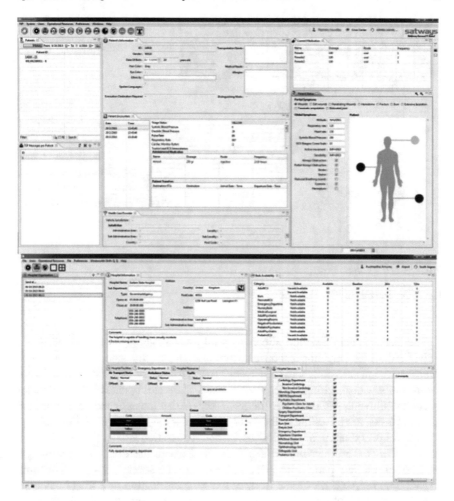

Fig. 2 The user-interface of the patient tracking module (left) and the hospital availability module (right)

2.2 INCIMOB—Mobile, Field Portable Command, Control and Patient Tracking Component

INCIMOB is defined as mobile extension for INCIMAG and is therefore only connected to this particular system. Each INCIMOB only communicates with the predefined INCIMAG instance; there is no direct connection between two INCIMOBs. INCIMOB itself comprises multiple functionalities. It provides features to handle status updates for field staff, send and receive situation reports, facilitates patient tracking, text messaging and it presents information sent from INCIMAG instances.

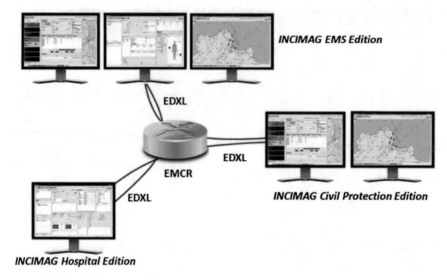

Fig. 3 An example of EMCR-usage

The status update function allows to send updates about own availability and activities during emergency response operations. On the other side, INCIMAG operators can assign each connected INCIMOB (user) to current incidents and tasks and therefore allow dispatching the INCIMOB units. The status updates follow a detailed workflow, allowing INCIMAG operators to see whether the assigned field units have accepted tasks, are travelling to the incident scene or having arrived at the scene.

The major function of INCIMOB is the registration and tracking of emergency patients. The registration of patients starts with the triage and the assignment of a unique patient-ID. INCIMOB supports several triage algorithms: START/JumpSTART, Sieve and Sort, or the simple selection of a triage colour. It also supports multiple possibilities of patient-IDs: electronic NFC-Tags, scanning of barcodes or simple manual entering of numbers. After the first registration of the patient, there are multiple ways of entering detailed patient data, vital signs and symptoms (see Fig. 4). Bluetooth enabled medical devices that support the Bluetooth Health Device Profile can be connected to INCIMOB. Vital Signs can be captured in real-time.

Patient data recorded by INCIMOB is then transferred to the connected INCIMAG, forwarded to the IMPRESS integration layer and processed by the PAT-EVO/LOGEVO components (see Fig. 5).

In addition, the application has a function to submit situational reports to INCIMAG. These reports can be enriched with a photo. This information gives the INCIMOB operators a better overview about the ongoing incident and the situation in the field.

The architecture of INCIMOB is designed as a set of modules; each of which provides a particular function to the system. To switch between the modules, the

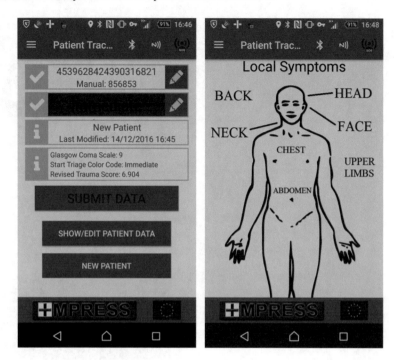

Fig. 4 INCOMOB, patient tracking

application uses a design pattern that is called 'navigation drawer'. This pattern is used in a wide range of mobile applications and is designed to be intuitive for persons that use smartphones in their everyday life.

2.3 Reference Semantic Model—Ensuring All Components and Users Understand Each Other

The Reference Semantic Model defines the IMPRESS Ontology. When constructing this ontology the objectives were interoperability, data harmonization and linked data provision. The IMPRESS Ontology design and implementation process follows the METHONTOLOGY [1] steps and is implemented in OWL (Web Ontology Language). The IMPRESS Ontology upper layer contains the following four main concepts:

- EOPHC (Extra Ordinary Public Health Challenge): The concept refers to the emergency events and incidents that take place and require response.
- Person: The concept refers to the human individual.
- Resource: The concept refers to anything that is used to support or help in the response during a health emergency.

Fig. 5 INCIMOB: patient
tracking recorded and sent to
INCIMAG

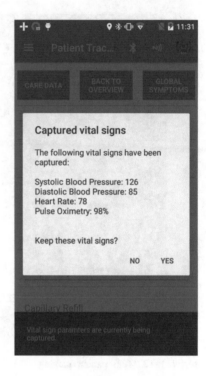

- Activity: The concept refers to any activity that takes place in order to reduce the impact of an emergency event.

The temporal aspects of the ontology have also been taken under consideration describing the evolution of the data through time. The data fact includes various sets of code lists described in SKOS (Simple Knowledge Organization System) as well as properties that are associated with user roles and geospatial data. The IMPRESS Ontology is further aligned with the TSO (Type-Based Service Ontology) standard.

While not a software component in its own right the Reference Semantic Model provides a well-defined ontology that provides a set of mutually defined concepts which in turn allow the software components of the IMPRESS system to interact in a mutually intelligible way.

2.4   The Message Bus—Inter-component Communication

The IMPRESS system uses Apache ActiveMQ to pass data and instructions between the individual components of the system. While this may be considered a standard piece of software and one that was not developed by or for the IMPRESS project it is mentioned here for completeness and reader comprehension.

2.5 Data Harmonization Component—Ensuring All Data Within the System Conforms to the Ontology

The Data Harmonization Component implements the data harmonization procedure that is required in order to harmonize the multidisciplinary and heterogeneous datasets of the IMPRESS Platform and provide a semantically homogenized view of the data. These data are provided as linked data to the rest of the IMPRESS Components. Thus, Data Harmonization is also responsible for the linking of the RDF (Resource Description Framework) data [2] with other third party linked data resources.

The Data Harmonization Component implements data harmonization using the IMPRESS Reference Semantic Model, which covers the respective domains of knowledge of the domain of health emergency management.

The main tasks that are realized by the Data Harmonization Component are the following:

1. Provide a real-time RDF view of the IMPRESS data stored in the WARSYS component
2. Provide access to RDF data views via a SPARQL endpoint
3. Execute SPARQL queries to RDF data and process the results, if necessary
4. Links the IMPRESS RDF data with specific linked data resources
5. Handle and serve the requests for data from the Message Bus.

Based on the above, Data Harmonization Component includes four main sub-components: the Mapping Generator which is responsible to generate the mappings between the database data and the RDF view, the RDF Viewer which provides access and exposes the RDF views, the Query Handler subcomponent which handles the requests for data and the Data Linker.

2.6 WARSYS—The Central Data Store Component

WARSYS is the central data warehousing and information storage component of the IMPRESS system. It provides the interfaces to import data from medical and logistics repositories (such as hospital information systems) and is used to store and view the extracted data and information generated inside the IMPRESS system.

WARSYS incorporates a set of Web Applications to provide the datasets it holds to its users (internal or external) together with human accessible mechanisms for data loading via a secure web interface. Once the user is granted access to the WARSYS, they are offered the ability to upload files as well as the ability to provide data through a web form. The data from this process is stored within WARSYS database and can be sent to other IMPRESS system components via the IMPRESS Messaging Bus in EDXL format.

In addition to the typical database-like capabilities WARSYS also provides a range of additional services to the IMPRESS system such as: a web interface to trigger the IMPRESS simulation process for training purposes, a report server where preconfigured reports are stored and served to users, an admin reporting Interface, a web service providing bed availability in XML format or in JSON, and a Lessons Learn Tool. The IMPRESS Lessons Learnt Tool (LLT) is a web interface within WARSYS where information regarding past incidents is managed. The LLT features an interface for domain experts to view, edit of mass casualty incident information, provide their opinion, attach past cases, comment and assign actions based on Key Performance Indicators (KPI) such as: time to arrival of the first ambulance on scene, time until the first ill/injured victim has been triaged in the field or the time the scene was declared safe.

3 Decision Support, Training and Public Engagement Components

3.1 PATEVO—Forecasting of Patient Needs and Physiological State

The main goal PATEVO is assisting health service professionals in the care of affected individuals and so providing patients with the most effective treatment possible even in case of resource shortage. PATEVO is able to work according to two different modalities: a MONO mode and a FULL mode. In the MONO approach, the physiological status of a victim is described by means of his/her Physiologic Point State (PPS), that is, a vector of values for the Physiological State Variable (PSV) levels and a vector for the related rates of worsening. The MONO mode can be used in a simulated crisis for the purpose of training or subcomponents testing: patients are generated by the system with their injuries and PSVs which remain unknown to the caregivers who must treat them and who can evaluate the patients only on the basis of the observed symptoms. The discrepancy between triage evaluation and the real patient status will help the trainees to develop better patient management skills.

In the FULL approach, instead, each patient is characterized by a Physiologic Distribution State (PDS) that is a collection of (numerically approximated) density functions, one for each of the levels of the physiological variables as well as for each of the relative rates of worsening. The FULL modality is characterized by the uncertainty (formalized by probability density functions) about the position of the victim in the Physiological State Variable space. This is the case, for example, when IMPRESS is used to deliver decision support during the actual course of a real crisis when the actual evolution of the patient's physiological state must be predicted. In a simulated situation, the user chooses a type of crisis event from the Event Library, a pre-compiled (updatable) library, including a series of possible incidents. Over time, the evolution is also determined by therapeutic manoeuvres (therapies), if any, delivered by the considered Assets. The system provides support to the decision maker suggesting what assets in what quantities are to be employed to restore or improve the PSV's values by delivering the appropriate therapies.

3.2   LOGEVO—Forecasting the Use and Availability of Resources

LOGEVO seeks to forecast of the evolution of the provision of resources to the hospital and to the field (Hospital Surge and similar) determining the time-curve of the amount of resources that can be provided to the system by exploiting the incremental capacity of the health structures involved in the crisis. By means of interconnections with other DSS components, in particular with PATEVO and the Recommendation Engine, LOGEVO allows the decision maker to inform their strategy for patient dispatch, and permits them to strike a balance between resource availability, resources deployed, expected resource provision, current and expected needs.

LOGEVO is composed of two libraries (Ambients and Assets) and of the mathematical formulations of the time evolution of the Asset provision. The Ambient Library contains the most important categories of structures delivering resources (human and material). Each Ambient delivers one or more important resources (called "asset"). The mathematical model predicting the evolution over time of the provision of a certain "asset" incorporates by means of its formulation, the dependency of surge capacity on the type and severity of the event. The function returns the prediction of the amount of each "asset", in correspondence of each ambient type, given the number of the people involved in the crisis scenario, and the level of the asset before the incident occurred. While LOGEVO returns a best-guess forecast about the theoretical ability of each involved health care facilities to provide increased levels of resources, this theoretical forecast is balanced, within IMPRESS, against the concrete limits on the effective utilization of such resources imposed by the context. So for example, while LOGEVO may predict that, under the current crisis severity conditions, a hospital of a given type could triple its offer of Operating Rooms within twelve hours, if one specific Hospital of that type has been put out of operation by the crisis itself (damaged by the earthquake, contaminated by toxicants etc.) other components of the IMPRESS system will exclude or limit recourse to that facility, and its theoretical increment of resource provisions will therefore be totally or partially ineffective.

3.3 Recommendation Engine—Distribution of Patients to Healthcare

The Recommendation Engine provides recommendations about optimal order of patient treatment and evacuation, the destination hospital and routes to the hospitals. The component determines these recommendations based on the patient status, the health status forecast, on available or forecasted resources (ambulance vehicles and hospital bed availabilities in different categories). This component can be used in every day incident patients dispatching and in case of mass casualty incidents.

The Recommendation Engine consists of the Distribution Service and the Optimisation Services: The distribution computation is based on routing on a street network that is loaded from a spatial database. Street network parameters used for the routing can be adjusted manually, to reflect situation specific traffic limitations as for example speed reductions or road blockings. For each patient, a transportation priority is determined according to the needs. Patients may need certain assets for medical treatment which determine the choice of the hospital. Also, the number of available assets and capabilities of each hospital may increase after declaring the MCI case and raising the alerting level of the hospital and decrease with their utilisation.

The optimisation computation tries to find an optimal patient transport order by an iterative procedure. It uses LOGEVO to determine the increasing hospital asset availabilities and PATEVO to assess the current health status of the patient, how the health status will evolve over time and how the treatment by means of the assets will influence the health status. As a next step, an initial priority and a list of needed assets for each patient is determined which is in turn used to compute the prehospital time of each patient via the distribution computation explained above. Based on the resulting prehospital time, PATEVO is used to check the health status at arrival time. If a patient would be dead or if other constraints are not met, several adjustments on the patient priorities are done and the computations are repeated. This is repeated several times to find optimal patient priorities based on a Greedy algorithm.

3.4 SORLOC—Biological Source Localisation in Time and Space

Epidemiological practice during the investigation of an outbreak of a disease can be greatly assisted by the location of the source of a disease. Answering such a question in a systematic, mathematical manner, when the source is unknown can permit further questions to be answered such as forecasting where further cases are likely to occur or what the total number of casualties are likely to be. To be of greatest value such estimates need to be made from a small number of cases early in the course of the outbreak so that public health officials may prioritize resources. Diseases however may not be limited to a single source or event. Examples where there may be multiple clusters within an outbreak could include: a legionella outbreak where multiple cooling towers or air conditioning units are responsible for the cases, a shipment of infected food distributed to a large number restaurants, schools, canteens etc. over a region or perhaps a terrorist incident involving multiple covert releases of a pathogen in a short period of time. SORLOC therefore has a single scope: determine a SOuRce LOCation. This determination has two facets, one temporal and one spatio-temporal [3].

SORLOC, when given cases of a known disease via the IMPRESS message bus, uses a Dirichlet Process framework to solve the source location problem and return the results of the calculation back via the message bus. Greater detail of how this process is performed is available in [4]. A human accessible frontend to this process has been integrated into INCIMAG so that that cases and casualties may be fed directly into SORLOC and the results returned directly to the operator so that action may be taken immediately.

3.5 IMPRESS Training Component

The IMPRESS Training Component provides a software platform that allows users to develop competences and skills (both in terms of domain processes and the use of the IMPRESS platform) relevant to the efficient and effective handling of health crisis incidents. The platform allows the implementation of the training process both in traditional (on site) or distance-learning modes.

The IMPRESS training suite consists of five different parts, according to the focused learner group addressed and the scope of the training. In particular:

Domain training: It covers the emergency preparedness and response of the health services to extra-ordinary public health challenges (EOPHC). A fundamental component of the Domain Training material is the Post-crisis evaluation and lessons learnt collection tool: this tool allows team members and partners to discuss successes during a health emergency incident, including unintended outcomes, and recommendations for others involved in similar future cases. It also allows stakeholders to discuss things that might have been done differently, to identify the root causes of problems that occurred, and to suggest ways to avoid those problems in future handling of such incidents.

IMPRESS platform training: It provides all trainees with a general description of the IMPRESS solution and the relative ICT platform including: aim, objectives, general characteristics, functions, relationship with existing business processes and how these are affected/improved through the use of the project solution.

IMPRESS modules training: It provides individual trainees with a basic description of the main IMPRESS functional modules (hardware and software).

IMPRESS functions training: It provides trainees with instructions on how to use the IMPRESS system and each of the IMPRESS software modules for the implementation of their business processes.

IMPRESS training for decision-makers: It provides high-level decision-makers with knowledge regarding how the IMPRESS platform can be adopted and strategically deployed by an organization.

The training platform complies with the General Guidelines of the Data Security Protocol developed for the IMPRESS project. Compliance is achieved through the use of a secure access mechanism which features system user/network protocol authentication mechanisms, data-loss prevention capabilities, explicit identification of user roles/corresponding access rights, and protection against attacks such as cross-site scripting, command injection, and buffer overflow.

3.6   INCIcrowd—Permitting and Channelling Public Assistance

INCIcrowd is a light version of INCIMOB and enables the public to receive alerts, submit observations and exchange resource offers and needs. Thus, it facilitates crowdsourcing.

INCIcrowd is connected to the IMPRESS system via a dedicated server and the IMPRESS message bus. The architecture is very similar to INCIMOB, as it uses the same framework. Nevertheless, the functionalities are much reduced compared to INCIMOB.

In order to reduce the barriers to use this application, INCIcrowd does not require a personal user account and handles data as anonymous as possible to preserve data privacy. Sending observations and receiving alerts is completely anonymous. Observations will be collected and bundled by the dedicated Crowdsourcing Server who forwards the observations to INCIMAG. There it is up to the operator to display and use the information for decision making.

The other way around, all alerts that are published by INCIMAG and marked as public are collected by the Crowdsourcing Server and forwarded to the INCIcrowd applications. INCIcrowd can be configured for several regions to display related alerts.

An additional feature of INCIcrowd is to publish and read offers and needs of citizens, without coordination by officials. It is designed as an electronic bulletin board and allows to search for items and to send private messages to the authors of offers or needs to negotiate the transfer.

3.7   Ethical Issues Addressed in IMPRESS System

One of the objectives of IMPRESS was to include crowd sourcing for sharing data with volunteers and affected people through the INCICrowd app. By definition, crowd sourcing combines the efforts of numerous self-identified volunteers or part-time workers, where each contributor, acting on their own initiative, adds a small contribution that combines with those of others to achieve a greater result. The issue of data protection in crowd sourcing initiatives was very important thus IMPRESS system has to address it. The disaster response work, data protection is a key prerequisite, therefore, IMPRESS Ethical Review Committee (ERC) have been pointed out to facilitate this need. Elaboration of crowd sourcing schemes in a way to respect privacy (including location and communication privacy) of persons reporting and collaborating; and to comply with data protection principles (with focus on grounding data processing on informed consent, purpose limitation, data minimization, accuracy of personal data, limited retention of data, information about the rights of participants).

Proper governance arrangements were essential to ensure that any risks to the lawful and ethical conduct of the project were identified and minimised so as not to compromise legal and ethical standards. IMPRESS has put in place such governance arrangements through the establishment of a Committee. The Committee's first step was the formulation of a Data Protection Framework to guidelines of IMPRESS research activities. A Privacy Impact Assessment (PIA) was conducted, focusing on identifying the impacts on privacy of any new project, technology, service or programme and, in consultation with stakeholders, taking remedial actions to avoid or mitigate any risks.

Specific attention has been paid to system and data security requirements. Data security had to be preserved through, encryption, and regulation of the transfer. Data controllers Researchers acting as data processors were responsible and accountable for complying with security requirements as provided by the law.

An important part of the work carried out in IMPRESS project, dealt with the OHS (Occupational safety and health) issues encountered by first responders. In particular, psychosocial and musculoskeletal problems are usual with first responders [5]. Heat stress, gait, CBRN contamination and ergonomic issues (among others) have a long-term impact on first responders' health. Working on shifts, long working hours, regular contact with traumatic scenes leave a long-term psychosocial trauma, usually expressed through stress related diseases. All this coupled with the fact that IMPRESS deals with emergency responders in the health sector creates an even more difficult mix of possible risks. As part of the IMPRESS risk assessment framework focusing on emergency responders was prepared and multi-hazard approach that was identified as a need in order to cover the wider spectrum of the emergency services.

4   Evaluation of IMPRESS DSS Solution

4.1   Methodology

In order to grasp the reactions of the stakeholders and collect useful feedback on evaluating both the IMPRESS solution and the prototype system, several information collection methods and techniques were used. The main methods included the use of end-session questionnaires, group discussions guided by specific key evaluation questions (KEQs) and personal interviews with experts, Scientific Advisory Group (SAG) members, event's participants and representatives of emergency organizations.

More specifically the consortium used consistently end-of-session questionnaires to collect immediate feedback about the IMPRESS solution or the demonstrated prototype during the project events including workshops, training, conferences, trial, exercise and demo days or SAG meetings.

IMPRESS KEQs have been arranged according 3 basic types of evaluation concerning a. the IMPRESS approach (concept), b. the project implementation and outcome (IMPRESS prototype) and c. the operational evaluation. The above questions have been defined following unstructured interviews of the consortium with members of the SAG of the project and are considered adequate to grasp the feeling and collect the feedback of the stakeholders involved in the evaluation of IMPRESS.

Finally, one of the most efficient techniques for collecting evaluation comments and feedback concerning the system were considered the interviews with stakeholders. These interviews held in order to present the approach of the project (IMPRESS solution) as well as for getting reactions concerning the demonstrated system capabilities.

5 IMPRESS Trials

The evaluation of IMPRESS systems was highly supported by the large number of events organized by the project consortium in cooperation with end user's organizations. Originally there were foreseen two pilots of IMPRESS, one in Italy (Palermo) and a second one as a cross-border event between Bulgaria and Greece. During the course of the project it was decided and organized an extra field test in Montenegro (Podgorica), which provided another opportunity to test and evaluate IMPRESS under realistic conditions and in cooperation with operational end users (NATO Rescue Forces). Each of these three trials of the IMPRESS prototype had slightly different objectives in terms of evaluation as follows.

5.1 IMPRESS Palermo Trial (Italy)

The first trial and real-life testing of IMPRESS prototype was implemented in context of a large-scale field exercise and took place in Palermo, Sicily (Fig. 6), organized cooperatively by the IMPRESS consortium and the Italian Civil Protection in 2016, June 7th. More than twenty (20) organizations and public services participated in the trial with approximately 500 people deployed in the various sites of the exercise. The health emergency exercise scenario involved a fire on-board a cargo ship moored in

Fig. 6 IMPRESS Palermo exercise

front of the Palermo harbor, in the sea, right in front of the Palermo promenade. The fire could not be extinguished and the captain and crew should abandon ship. Due to Northeasterly winds the plume of toxic substances released in the fire reached the densely populated area of Kalsa District. Several victims were reported which needed medical attention and transportation to nearby hospitals; essentially triggering a mass casualty emergency operation.

IMPRESS was involved in the exercise as the back-end system coordinating the collection of field data, making medical and emergency assessments and managing the flow of information and data among the involved services during the response phase of the crisis management scenario. For this purpose, a number of INCIMAG (5) instances have been deployed and installed in the Regional Department of Civil Protection, the Prefecture of Palermo, the EMS SEUS 118, the Coast Guard facilities in the Port of Palermo and the Hospital Buccheri la Ferla. Furthermore, a number of INCIMOB (20) devices were assigned to EMS resources (ambulance crews) of Palermo Health District, Italian Red Cross, Civil Protection field personnel and volunteers, Fire Service SAR teams, Italian National Police and Carabinieri squads through the respective INCIMAGs.

5.2 IMPRESS Podgorica Trial (Montenegro)

The second trial was also real-life testing of the mobile part of the IMPRESS system as part of the Consequence Management Exercise CRNA GORA 2016 of NATO. A team from Crisis Management and Disaster Response Centre of Excellence (CMDR COE) and the IMPRESS project partners from INTRASOFT, IVI, IICT-BAS and ECOMED were operating on the field transmitting in real time data from Podgorica field test exercise to the desktop application located in Athens. The IMPRESS system was tested in real time cross-border conditions with sparse network coverage. The exercise was conducted by the NATO's Euro-Atlantic Disaster Response Coordination Centre (EADRCC) and the Ministry of Interior of Montenegro and took place from 31 October to 4 November 2016 in Podgorica.

INCIMOB and INCIMAG components of the IMPRESS system were tested there. The tools were used to collect data from injured people on the field and dispatch center was supporting from Athens all decision actions on the field teams. More than 680 participants from 10 Partner and 7 Allied nations (Albania, Armenia, Austria, Azerbaijan, Belgium, Bosnia and Herzegovina, Croatia, Finland, FYROM, Georgia, Israel, Romania Serbia, Slovenia, Spain, Sweden, Ukraine and USA) practiced in the exercise. Disaster response mechanisms and exercise capabilities were tested to work together effectively in emergency situations. The exercise also contributed to strengthen the affected nation's capacity to effectively coordinate crisis response operations.

5.3 Greek-Bulgarian Cross-Border Table Top Exercise (Sofia, BG)

The final test of the IMPRESS solution and the system's prototype was held in Sofia as a tabletop exercise (TTX) (Fig. 7), performed around a cross-border scenario involving a strong earthquake and flood emergency. The scenario was related to a large-scale incident in South Bulgaria due to the combined earthquake and flood occurrence, necessitating cross-border emergency cooperation with Greece through ERCC (Emergency Response Coordination Center, EC).

For the needs of the TTX a number of INCIMAGs, INCIMOB and INCICROWD devices were used by operators from emergency services, hospitals' personnel, volunteers and civil protection. The IMPRESS resources were 6 INCIMAGs for the Bulgarian part, 5 INCIMAGs for the Greek part and 1 INCIMAG as ERCC. In addition, 15 INCIMOB and 5 INCICROWD devices were distributed to the actors that played the field data collection, situation reporting and observation posting role. Taking into account that each INCIMAG had two operators a number of approximately 50 stakeholders experienced the use and operation of the IMPRESS resources and they were able to provide proven evaluation inputs and comments. The end users who operated the IMPRESS resources were trained in physical meetings in Sofia

Fig. 7 IMPRESS Greece-Bulgarian TTX

and Thessaloniki one month before the TTX, they have been provided access to the online training platform of IMPRESS and they had half day training in Sofia the day before the exercise.

6 Feedback and Lessons Learned

The final conclusions from the evaluation of the IMPRESS solution and the prototype system were positive. The flexibility of the system architecture and the modular design which allows the system to expand has been considered ideal to allow the system being adapted to diverse organizational structures. Also, the ability of the system to fit both to routine operation of emergency medical services and to multi-casualty incidents and multi-agency coordination during emergencies was considered a significant feature. The stakeholders appreciated the design characteristics of the system, which allow for data privacy despite the fact that this issue is questioned by practitioners in case of large disasters and emergencies.

Moreover, it has to be noticed the deployment of the prototype system within few days in different countries as well as the operation of the system prototype by operational personnel following brief training. Both these facts were evaluated positively by end users. Positively were also evaluated the capability of the system to address the requirements of multiple stakeholders involved in disaster response providing an end-to-end solution for managing emergency and medical information during disasters. Furthermore, the integration of IMPRESS with legacy systems (HIMS/FILOLAOS, Greece/EKEPY-National Health Coordination And Command Center) it was appreciated by end users since it is always desirable however in most cases impossible to demonstrate such integration in context of R&D projects.

Further to the above feedback, also significant gaps between the operational needs of medical response and the realm of large disasters from one side; and the capabilities of the technology and the potential of its application in this field, have derived. As regards the trials and exercises the most significant lesson learned is that more time is needed for the practitioners to perceive and reflect to the use of innovative technological solutions to their operational practice. Practitioners become more interested and positive in case they have the possibility to put their hands on the system rather than hear or watch its capabilities.

Finally, concerning the reality of the medical emergencies during large disasters or Mass Casualty Incidents (MCIs) the lesson learned is all about time. The medical staff on the scene has very limited time to type data and the field conditions are very harsh to be able to use and watch devices. Thus, data input methods should be improved accordingly.

Acknowledgements This work has been partially supported by the EC Research Executive Agency 7th Framework Programme, (SEC-2013.4.1-4) under grant number: FP7-SEC-2013-608078-IMproving Preparedness and Response of HEalth Services in major criseS (IMPRESS), the UK NIH Health Protection Research Unit in Emergency Preparedness and Response and the Bulgarian National Science Fund project number DN 12/5 called: Efficient Stochastic Methods and Algorithms for Large-Scale Computational Problems.

References

1. Gomez-Perez, A., Fernández-López, M., Corcho, O.: METHONTOLOGY: From Ontological Art Towards Ontological Engineering. Ontological Engineering: with examples from the areas of Knowledge Management, e-Commerce and the Semantic Web. Springer Science & Business Media (2006)
2. Resource Description Framework (RDF), https://www.w3.org/RDF/
3. Bailey, N.T. (ed.): The Biomathematics of Malaria. The Biomathematics of Diseases: 1 (1982)
4. Prentice, R.L., Pyke, R.: Logistic disease incidence models and case-control studies. Biometrika, pp. 403–411 (1979)
5. Thompson, J., et al.: Risks to emergency medical responders at terrorist incidents: a narrative review of the medical literature. Crit. Care **18**, 521 (2014). PMC. Web. 25 Jan. 2018

The New Approach for Dynamic Optimization with Variability Constraints

Paweł Drąg and Krystyn Styczeń

Abstract In this work a new optimization approach for processes modeled by differential-algebraic equations with variability constraints was presented. The designed procedure was based on the modified direct shooting method, which can transform the dynamic optimization problem into a large-scale nonlinear optimization task (NLP). The first-order KKT optimality conditions with complementarity constraints were obtained. Finally, to solve the optimality conditions with the complementarity constraints, the solution procedure combining SQP algorithm with the filter approach as a globalization procedure was designed. The efficiency of the presented methodology was tested on a production process in chemical engineering.

Keywords Dynamic optimization · Differential-algebraic equations · Variability constraints · Filter method · Complementarity constraints

1 Introduction

Optimization of the production processes is a task of a high importance. Especially, complexity of the mathematical models of the considered industrial systems is still growing. Therefore, the research on efficient optimization procedures is directly motivated by the observed reality [5, 20, 23, 27].

Some successful applications of the optimization methods in science and technology have been recently reported, especially in such areas as mechanical engineering, bioengineering and medicine. Pandelidis and Anisimov [25] presented an optimization approach of the cross-flow Maisotsenko cycle indirect evaporative air cooler. The optimization according to chosen five influence factors was performed with

P. Drąg (✉) · K. Styczeń
Department of Control Systems and Mechatronics, Wrocław University of Science and Technology, Janiszewskiego 11/17, 50-370 Wrocław, Poland
e-mail: pawel.drag@pwr.edu.pl

K. Styczeń
e-mail: krystyn.styczen@pwr.edu.pl

© Springer Nature Switzerland AG 2019
S. Fidanova (ed.), *Recent Advances in Computational Optimization*,
Studies in Computational Intelligence 795,
https://doi.org/10.1007/978-3-319-99648-6_3

two investigated methods: single-parameter, as well as multi-parameter compromise method. Moreover, recently a new approach for indoor air quality assessment has been designed [21]. Augspurger et al. [1] proposed a new strategy for the optimization of latent heat thermal storage device. In biotechnology, specified methabolic fluxes have been used to control and optimize the chemicals production [29]. Moreover, some of the other recently designed approaches have been applied in medicine to optimize the therapy [2, 26].

The presented work is concentrated on the production systems, which can be described by three specific types of constraints:

1. Differential-algebraic constraints (DAEs)
   Differential-algebraic constraints represent two main relations observed in the production systems: dynamics and conservation laws. It means, that the process can be described by classical ordinary differential equations (ODEs), extended by the additional algebraic formulations. Then, mass and energy conservation laws, as well as the algebraic Kirchhoff's relations, can be incorporated directly into the model. Therefore, the obtained mathematical model can be consisted on the physical laws in a commonly known formulation [6, 8].

2. The variability constraints
   The variability constraints for process optimization task have been introduced in [10]. Then, the presented formulation has been successfully applied for solving some highly nonlinear models consisted on the differential-algebraic equations [11–14]. The variability constraints have been introduced into nonlinear systems simulations as a general solution procedure. They enable us to preserve the process variability in an assumed range. Therefore, an approximated process solution can be always obtained. Finally, the approximated solution can be improved by the problem-specialized algorithms.

3. The complementarity constraints
   In mathematical programming the complementarity constraints can be used to represent inclusive ORs relation. Therefore, they have important modeling applications in such areas as physics (transition between static and kinetic friction), thermodynamics (phase equilibrium models) or economics (market equilibrium state) [9]. Moreover, the complementarity constraints can be applied to model switches instant in the designed mathematical models. The offer an alternative to mixed integer programming in nonconvex optimization tasks. A general form mathematical programming with equilibrium constraints (MPEC)

$$\min \quad f(\mathbf{x}) \tag{1}$$

subject to

$$0 \leq x_1 \perp x_2 \geq 0 \tag{2}$$

indicates, that either $x_1 = 0$ or $x_2 = 0$ (or both) and x_1, x_2 are non-negative. Moreover, the operator $\perp$ denotes x_1 completnes x_2. The direct form of the complementarity constraints $x_1 \cdot x_2 = 0$ violates NLP constraint qualifications at the solution point. Moreover, the constraints qualifications indicated, that the Lagrange multipliers are unique and bounded. The MPEC task can be reformulated and solved with a standard NLP solver [4]. In this work a strategy based on a penalty formulation was applied [3]. This approach enables us to introduce the complementarity constraints into to the objective function with an exact penalty function and a suitable penalty parameter

$$\min \quad f(\mathbf{x}) + \rho x_1^T x_2 \tag{3}$$

subject to

$$x_1, x_2 \geq 0 \tag{4}$$

The new approach for solving process optimization task subject to the technological constraints was based on the filter method. The idea of the filter method has been introduced by Fletcher and Leyffer [18]. Then some modifications of the filter approach have been successfully applied for solving large-scale nonlinear optimization problems. Wang et al. [28] designed an interior-point algorithm with new non-monotone line search filter method for nonlinear constrained programming. Echebest et al. [16] investigated an inexact restoration derivative-free filter method. Global convergence of a derivative-free inexact restoration filter algorithm for nonlinear programming has been discussed in [17]. Finally, a QP-free algorithm without a penalty function or a filter for nonlinear general-constrained optimization has been presented in [22].

The article is constructed as follows. In Sect. 2 the process optimization task was presented, as well as the nonlinear optimization problem was formulated. Then, the solution procedure was designed in Sect. 3. The results of numerical simulations were presented and discussed in Sect. 4. Finally, in Sect. 5, the obtained results were summarized.

2 Process Optimization Task

In our work it was assumed, that physical relations of the considered production process can be appropriately described by system of the differential-algebraic equations.

Assumption 1 The considered process can by described by the DAEs system of the following form

$$\dot{\mathbf{y}}(t) = \mathbf{F}(\mathbf{y}(t), \mathbf{z}(t), \mathbf{u}(t), \mathbf{p}, t)$$
$$0 = \mathbf{G}(\mathbf{y}(t), \mathbf{z}(t), \mathbf{u}(t), \mathbf{p}, t) \tag{5}$$

where $\mathbf{y}(t) \in \mathcal{R}^{n_y}$ denotes a differential state trajectory, $\mathbf{z}(t) \in \mathcal{R}^{n_z}$ is an algebraic state trajectory, $\mathbf{u}(t) \in \mathcal{R}^{n_u}$ is a control function, $\mathbf{p} \in \mathcal{R}^{n_p}$ is a global parameters vector constant in the time. The independent variable, eg. time, is denoted by $t \in \mathcal{R}$. The Jacobian matrix

$$\mathbf{G_z} = \frac{\partial \mathbf{G}}{\partial \mathbf{z}} = \frac{\partial G_i}{\partial z_j}, \quad i, j = 1, \dots, n_z \tag{6}$$

is invertible. Therefore, the index of DAEs system is not greater than 1. Moreover, two vector-valued functions are considered

$$\mathbf{F} : \mathcal{R}^{n_y} \times \mathcal{R}^{n_z} \times \mathcal{R}^{n_u} \times \mathcal{R}^{n_p} \times \mathcal{R} \to \mathcal{R}^{n_y} \tag{7}$$

$$\mathbf{G} : \mathcal{R}^{n_y} \times \mathcal{R}^{n_z} \times \mathcal{R}^{n_u} \times \mathcal{R}^{n_p} \times \mathcal{R} \to \mathcal{R}^{n_z}. \tag{8}$$

Then, according to the technological or numerical purposes, the variability constraints can be introduced.

Assumption 2 The variability constraints take the following form

$$\dot{\mathbf{y}}(t) \leq v_{\mathbf{y}} \tag{9}$$

where $v_{\mathbf{y}} \in \mathcal{R}^{n_y}$.

To solve the process optimization task with the differential-algebraic, as well as variability constraints, the direct shooting approach was proposed. The presented methodology was composed on the multiple shooting method with the direct shooting transcription approach. According to the multiple shooting method, the considered process duration time

$$t \in [t_0 \quad t_F] \tag{10}$$

was divided into assumed N subdomains as follows

$$t_0 = t_0^1 < t_F^1 = t_0^2 < t_F^2 = \cdots = t_0^N < t_F^N = t_F. \tag{11}$$

and in general

$$t^i \in [t_0^i \quad t_F^i], \quad i = 1, \dots, N. \tag{12}$$

Therefore, the process in each subinterval can be considered independently. Then, the mathematical model of the process was partially parameterized. In the classical approach the initial values of the differential state trajectories, as well as the control functions, were treated as the additional decision variables. Moreover, it was assumed, that the process dynamics in each subinterval is constant. In this work, this approach was extended, that the left-hand side of the dynamical equations is treated as decision variables also. Finally, in each stage the following parameterized model of the process was considered

$$\mathbf{x}_{\dot{\mathbf{y}}^i}(t^i) = \widetilde{F}^i(\mathbf{x}_{\mathbf{y}^i}, \mathbf{x}_{\mathbf{z}^i}, \mathbf{x}_{\mathbf{u}^i}, \mathbf{p}, t_0^i)$$
$$0 = \widetilde{G}^i(\mathbf{x}_{\mathbf{y}^i}, \mathbf{x}_{\mathbf{z}^i}, \mathbf{x}_{\mathbf{u}^i}, \mathbf{p}, t_0^i) \tag{13}$$

$$\mathbf{x}_{\dot{\mathbf{y}}^i} = \dot{\mathbf{y}}(t_0^i) \tag{14}$$

$$\mathbf{x}_{\mathbf{y}^i} = \mathbf{y}(t_0^i) \tag{15}$$

$$\mathbf{x}_{\mathbf{z}^i} = \mathbf{z}(t_0^i) \tag{16}$$

$$\mathbf{x}_{\mathbf{u}^i} = \mathbf{u}(t_0^i) \tag{17}$$

$$\mathbf{x}_{\mathbf{p}} = \mathbf{p} \tag{18}$$

and

$$\widetilde{F}^i : \mathcal{R}^{n_{y^i}} \times \mathcal{R}^{n_{z^i}} \times \mathcal{R}^{n_{u^i}} \times \mathcal{R}^{n_{\mathbf{p}}} \times \mathcal{R} \to \mathcal{R}^{n_{y^i}} \tag{19}$$

$$\widetilde{G}^i : \mathcal{R}^{n_{y^i}} \times \mathcal{R}^{n_{z^i}} \times \mathcal{R}^{n_{u^i}} \times \mathcal{R}^{n_{\mathbf{p}}} \times \mathcal{R} \to \mathcal{R}^{n_{z^i}} \tag{20}$$

The applied extended parameterization approach enabled us to define the vector of unknowns variables. The i-th stage was connected with the following vector of unknown variables

$$\mathbf{x}^i = [\mathbf{x}_{\dot{\mathbf{y}}^i} \quad \mathbf{x}_{\mathbf{y}^i} \quad \mathbf{x}_{\mathbf{z}^i} \quad \mathbf{x}_{\mathbf{u}^i}]. \tag{21}$$

Therefore, the matrix of the decision variables can be constructed

$$\mathbf{X} = \begin{bmatrix} \mathbf{x}_{\dot{\mathbf{y}}^1} & \mathbf{x}_{\mathbf{y}^1} & \mathbf{x}_{\mathbf{z}^1} & \mathbf{x}_{\mathbf{u}^1} \\ \mathbf{x}_{\dot{\mathbf{y}}^2} & \mathbf{x}_{\mathbf{y}^2} & \mathbf{x}_{\mathbf{z}^2} & \mathbf{x}_{\mathbf{u}^2} \\ \vdots & \vdots & \vdots & \vdots \\ \mathbf{x}_{\dot{\mathbf{y}}^N} & \mathbf{x}_{\mathbf{y}^N} & \mathbf{x}_{\mathbf{z}^N} & \mathbf{x}_{\mathbf{u}^N} \end{bmatrix}. \tag{22}$$

Then, the differential-algebraic constraints take the following form

$$\widetilde{F}^i(\mathbf{x}_{\mathbf{y}^i}, \mathbf{x}_{\mathbf{z}^i}, \mathbf{x}_{\mathbf{u}^i}, \mathbf{p}, t_0^i) - \mathbf{x}_{\dot{\mathbf{y}}^i} = 0$$
$$\widetilde{G}^i(\mathbf{x}_{\mathbf{y}^i}, \mathbf{x}_{\mathbf{z}^i}, \mathbf{x}_{\mathbf{u}^i}, \mathbf{p}, t_0^i) = 0 \tag{23}$$

for $i = 1, \ldots, N$. Therefore, the process constraints can be reduced

$$\widetilde{\mathbf{F}}(\mathbf{X}) = 0$$

$$\widetilde{\mathbf{G}}(\mathbf{X}) = 0 \tag{24}$$

The presented modified direct shooting approach enabled us to consider the variability constraints

$$\dot{y}^L \leq \dot{\mathbf{y}}(t) \leq \dot{y}^U \tag{25}$$

directly by

$$\mathbf{x}_{\dot{\mathbf{y}}}^L \leq \mathbf{x}_{\dot{\mathbf{y}}} \leq \mathbf{x}_{\dot{\mathbf{y}}}^U \tag{26}$$

where L and U denote the lower and upper bounds, respectively. Finally, the appropriate nonlinear optimization problem with a pointwise-continuous differential-algebraic constraints can be considered

$$\min_{\mathbf{X}} \quad f(\mathbf{X}) \tag{27}$$

subject to

$$\widetilde{\mathbf{F}}(\mathbf{X}) = 0$$

$$\widetilde{\mathbf{G}}(\mathbf{X}) = 0$$

$$\mathbf{X}_{\dot{\mathbf{y}}} \leq \mathcal{V}(\mathbf{X}) \tag{28}$$

$$\mathbf{X}^L \leq \mathbf{X} \leq \mathbf{X}^U$$

3 The Solution Procedure

As the results of the previous section, the optimization task with both equality and inequality constraints, as well as lower- and upper-bounds was obtained [Eqs. (27) and (28)]. The such constructed optimization problem can be solved using a sequential solution approach [15]. However, the inequality constraints on the state variables and their variability, can cause potential difficulties with derivative discontinuities. Therefore, the formulated optimization task were replaced by their first-order KKT optimality conditions, resulting in a single-level mathematical program with complementarity constraints (MPCC) [19]

$$\begin{aligned}
\nabla_{\mathbf{X}} f(\mathbf{X}) - \nabla_{\mathbf{X}}\widetilde{\mathbf{F}(\mathbf{X})}\lambda_{\widetilde{\mathbf{F}}} - \nabla_{\mathbf{X}}\widetilde{\mathbf{G}(\mathbf{X})}\lambda_{\widetilde{\mathbf{G}}} - \\
\mu_1 \nabla_{\mathbf{X}}(\mathbf{X}_{\dot{\mathbf{y}}} - \mathcal{V}(\mathbf{X})) - \\
\mu_2 \nabla_{\mathbf{X}}(\mathbf{X}^L - \mathbf{X}) - \\
\mu_3 \nabla_{\mathbf{X}}(\mathbf{X} - \mathbf{X}^U) = 0
\end{aligned}$$

$$\widetilde{\mathbf{F}(\mathbf{X})} = 0$$

$$\widetilde{\mathbf{G}(\mathbf{X})} = 0$$

$$\mu_1(\mathbf{X}_{\dot{\mathbf{y}}} - \mathcal{V}(\mathbf{X})) = 0 \tag{29}$$

$$\mu_2(\mathbf{X}^L - \mathbf{X}) = 0$$

$$\mu_3(\mathbf{X} - \mathbf{X}^U) = 0$$

$$\mu_1, \mu_2, \mu_3 \geq 0$$

The presence of the complementarity constraints may pose difficulties for nonlinear programming (NLP) procedures. Therefore, some MPCC problem reformulation strategies have been proposed. Here, an exact penalty method was applied, in which the complementarity constraints are removed to the objective function

$$f(\mathbf{X}) + \rho\Big(\mu_1(\mathbf{X}_{\dot{\mathbf{y}}} - \mathcal{V}(\mathbf{X})) + \mu_2(\mathbf{X}^L - \mathbf{X}) + \mu_3(\mathbf{X} - \mathbf{X}^U)\Big) \tag{30}$$

The reformulated nonlinear optimization task can be solved by the SQP filter algorithm. The idea of the filter approach is to consider the NLP problem as a bi-objective optimization task. Then, two conflicting purposes are considered: minimizing both the objective function

$$\min_{\mathbf{X}} \quad f(\mathbf{X}), \tag{31}$$

as well as the constraints violation

$$\mathbf{c}_{eq}(\mathbf{X}) = 0$$

$$\mathbf{c}_{in}(\mathbf{X}) \leq 0 \tag{32}$$

considered as

$$\min_{\mathbf{X}} \quad h(\mathbf{X}) = \sum_{i=1}^{n_{c_{eq}}} |c_{eq,i}(\mathbf{X})| + \sum_{i=1}^{n_{c_{in}}} |c_{in,i}(\mathbf{X})|^- \tag{33}$$

$$|c_{in,i}(\mathbf{X})|^- = \begin{cases} 0, & c_{in,i}(\mathbf{X}) \le 0 \\ \\ |c_{in,i}(\mathbf{X})|, & \text{otherwise} \end{cases} \tag{34}$$

The filter method accept a new trial step $\mathbf{X}^k$ as a new iterate if the pair $(f(\mathbf{X}^k), h(\mathbf{X}^k))$ is not dominated by a previous pair $(f(\mathbf{X}^l), h(\mathbf{X}^l))$ generated by the optimization algorithm, according to the Definition 1 [24].

Definition 1

1. A pair $(f(\mathbf{X}^k), h(\mathbf{X}^k))$ is said to dominate another pair $(f(\mathbf{X}^l), h(\mathbf{X}^l))$ if

$$f(\mathbf{X}^k) \le f(\mathbf{X}^l) \quad \text{and} \quad h(\mathbf{X}^k) \le h(\mathbf{X}^l). \tag{35}$$

2. A filter is a list of pairs $(f(\mathbf{X}^l), h(\mathbf{X}^l))$ such that no pair dominates any other.
3. An iterate $\mathbf{X}^k$ is said to be acceptable to the filter if $(f(\mathbf{X}^k), h(\mathbf{X}^k))$ is not dominated by any pair in the filter.

In the proposed methodology following two inputs of the filter are considered $\widetilde{f}(\mathbf{X})$ = squared 1st KKT condition with the modified objective function Eq. (30)

$$\widetilde{h}(x) = \sum_{i=1}^{N} |\widetilde{F}^i(\mathbf{X})| + \sum_{i=1}^{N} |\widetilde{G}^i(\mathbf{X})| \tag{36}$$

Finally, the modified acceptability condition were introduced by a sufficient decrease coefficient β. A trial iterate $\mathbf{X}^k$ is acceptable to the filter if, for all pairs $(\widetilde{f}(\mathbf{X}^j), \widetilde{h}(\mathbf{X}^j))$ in the filter, the following condition is satisfied

$$\widetilde{f}(\mathbf{X}^k) \le \widetilde{f}(\mathbf{X}^j) - \beta \widetilde{f}(\mathbf{X}^j) \quad \text{or} \quad \widetilde{h}(\mathbf{X}^k) \le \widetilde{h}(\mathbf{X}^j) - \beta \widetilde{h}(\mathbf{X}^j) \tag{37}$$

for $\beta \in (0, 1)$.

4   The Application in Process Industry

In the presented case study, a mathematical model of CSTR reactor was considered [7]

$$\min_{u(t)} \quad \mathcal{J} = \int_0^{0.78} \left(y_1^2(t) + x_2^2(t) + Ru^2(t) \right) \tag{38}$$

subject to

$$\dot{y}_1(t) = -2(y_1(t) + 25) + (y_2(t) + 0.5) \exp\left(\frac{25y_1(t)}{y_1(t) + 2}\right) - (y_1(t) + 0.25)u(t) \tag{39}$$

$$\dot{y}_2(t) = 0.5 - y_2(t) - (y_2(t) + 0.5) \exp\left(\frac{25y_1(t)}{y_1(t) + 2}\right) \tag{40}$$

$$\mathbf{y}(0) = \begin{bmatrix} y_1(0) \\ y_2(0) \end{bmatrix} = \begin{bmatrix} 0.05 \\ 0.00 \end{bmatrix} \tag{41}$$

$$u(t) \geq 0.0 \tag{42}$$

$$y_2 \geq -0.06 \tag{43}$$

where $y_1(t)$ is the deviation from steady-state temperature, $y_2(t)$ is the deviation from steady-state concentration. Finally, $u(t)$ is the normalized effect of coolant flow on the reaction. Moreover, the variability constraints were introduced

$$|y_1(t)| \leq 0.1 \quad [1/s] \tag{44}$$

$$|y_1(t)| \leq 0.2 \quad [1/s] \tag{45}$$

To solve the presented dynamical optimization task, the modified direct shooting approach was used. The filter algorithm was implemented in *Mathematica* einvronment and itertively applied two subroutines NSolve and NDSolve. As a main results, the final value of the objective function was equal to 0.027 after 150 sec. Intel(R) Core(TM) i7-4510U CPU computations time, 2.00 GHz, 8 GB RAM. The state profiles were represented by a piecewise linear approximation at 7 equidistant grid points (Figs. 1 and 2). The control profile was approximated by piecewise constant approximation (Fig. 3).

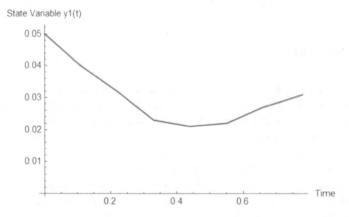

Fig. 1 State variable $y_1(t)$ profile for CSTR reactor

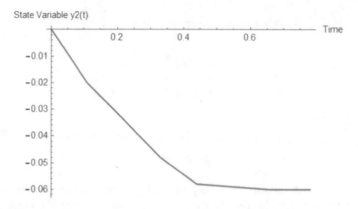

Fig. 2 State variable $y_2(t)$ profile for CSTR reactor

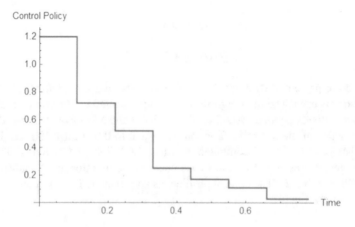

Fig. 3 Control variable profiles for CSTR reactor

5   Conclusion

In the presented work, the dynamic process optimization task was considered. The designed algorithm was based on the extended direct shooting approach. Moreover, the variability constraints were incorporated. The inequality form of the variability constraints resulted in difficulties in the obtained nonlinear optimization task. Moreover, the obtained first-order KKT optimality conditions take a MPEC form. Therefore, the complementarity constraints were removed into the objective function with the penalty term. Finally, to solve such constructed task, the filter SQP algorithm was implemented. The designed procedure was investigated on the modified CSTR reactor dynamic optimization task.

Acknowledgements One of the co-authors, Paweł Drąg, received financial support in the framework of "Młoda Kadra"- "Young Staff" 0402/0109/17 at Wrocław University of Science and Technology.

References

1. Augspurger, M., Choi, K.K., Udaykumar, H.S.: Optimizing fin design for a PCM-based thermal storage device using dynamic Kriging. Int. J. Heat Mass Transf. **121**, 290–308 (2018). https://doi.org/10.1016/j.ijheatmasstransfer.2017.12.143
2. Babaei N., Salamci M.U.: Controller design for personalized drugadministration in cancer therapy: successive approximation approach. Optimal Control Appl. Methods **138** (2017). https://doi.org/10.1002/oca.2372
3. Baumrucker, B.T., Biegler, L.T.: MPEC strategies for cost optimization of pipeline operations. Comput. Chem. Eng. **34**(6), 900–913 (2010). https://doi.org/10.1016/j.compchemeng.2009.07.012
4. Biegler L.T.: Nonlinear Programming: Concepts, Algorithms, and Applications to Chemical Processes. Society for Industrial and Applied Mathematics, Philadelphia (2010). https://doi.org/10.1137/1.9780898719383
5. Biegler L.T.: Advanced optimization strategies for integrated dynamic process operations. Comput. Chem. Eng. (2017) (in press). https://doi.org/10.1016/j.compchemeng.2017.10.016
6. Biegler L.T., Campbell S.L, Mehrmann V.: Control and Optimization with Differential-Algebraic Constraints. Society for Industrial and Applied Mathematics, Philadelphia (2012). https://doi.org/10.1137/9781611972252.fm
7. Bloss, K.F., Biegler, L.T., Schiesser, W.E.: Dynamic process optimization through adjoint formulations and constraint aggregation. Ind. Eng. Chem. Res. **38**(2), 421–432 (1999). https://doi.org/10.1021/ie9804733
8. Cao, Y., Li, S., Petzold, L., Serban, R.: Adjoint sensitivity analysis for differential-algebraic equations: The adjoint DAE system and its numerical solution. SIAM J. Sci. Comput. **24**(3), 1076–1089 (2003). https://doi.org/10.1137/S1064827501380630
9. Dowling, A.W., Balwani, C., Gao, Q., Biegler, L.T.: Optimization of sub-ambient separation systems with embedded cubic equation of state thermodynamic models and complementarity constraints. Comput. Chem. Eng. **81**, 323–343 (2015). https://doi.org/10.1016/j.compchemeng.2015.04.038
10. Drąg, P.: Algorytmy sterowania wielostadialnymi procesami deskryptorowymi. Akademicka Oficyna Wydawnicza EXIT, Warsaw (2016)
11. Drąg P., Styczeń K.: A general optimization-based approach for thermal processes modeling. In: Proceedings of the 2017 Federated Conference on Computer Science and Information Systems: September 3–6, 2017, pp. 1347–1352. Prague, Czech Republic (2017). https://doi.org/10.15439/2017F458
12. Drąg P., Styczeń K.: A new optimization-based approach for aircraft landing in the presence of windshear. In: Communication Papers of the 2017 Federated Conference on Computer Science and Information Systems: September 3–6, 2017, pp. 83–88. Prague, Czech Republic (2017). https://doi.org/10.15439/2017F436
13. Drąg P., Styczeń K.: Process Control with the Variability Constraints. In: Fidanova S. (ed.) Recent Advances in Computational Optimization. Studies in Computational Intelligence, vol. 717, pp. 41–51. Springer (2018). https://doi.org/10.1007/978-3-319-59861-1_3
14. Drąg P., Styczeń K.: The variability constraints in simulation of index-2 differential-algebraic processes. In: Nivitzka, V., Korecko, S., Szakal, A. (eds.) INFORMATICS 2017: 2017 IEEE 14th International Scientific Conference on Informatics, November 14–16, 2017, Proceedings, pp. 80–86. Poprad, Slovakia (2017)

15. Drąg P., Styczeń K., Kwiatkowska M., Szczurek, A.: A review on the direct and indirect methods for solving optimal control problems with differential-algebraic constraints. In: Recent Advances in Computational Optimization, vol. 2016, pp. 91–105. Springer (2016). https://doi.org/10.1007/978-3-319-21133-6_6
16. Echebest, N., Schuverdt, M.L., Vignau, R.P.: An inexact restoration derivative-free filter method for nonlinear programming. Comput. Appl. Math. **36**, 693–718 (2017). https://doi.org/10.1007/s40314-015-0253-0
17. Ferreira, P.S., Karas, E.W., Sachine, M., Sobral, F.N.C.: Global convergence of a derivative-free inexact restoration filter algorithm for nonlinear programming. Optimization **66**, 271–292 (2017). https://doi.org/10.1080/02331934.2016.1263629
18. Fletcher, R., Leyffer, S.: Nonlinear programming without a penalty function. Math. Program. Ser. B **91**, 239–269 (2002). https://doi.org/10.1007/s101070100244
19. Jamaludin, M.Z., Li, H., Swartz, C.L.E.: The utilization of closed-loop prediction for dynamic real-time optimization. Can. J. Chem. Eng. **95**, 1968–1978 (2017). https://doi.org/10.1002/cjce.22927
20. Koller, R.W., Ricardez-Sandoval, L.A., Biegler, L.T.: Stochastic back-off algorithm for simultaneous design, control and scheduling of multi-product systems under uncertainty. AIChE J. (2018) (in press). https://doi.org/10.1002/aic.16092
21. Kwiatkowska M., Szczurek A., Drąg P.: Zastosowanie równań różniczkowo-algebraicznych do predykcji zmian parametrw powietrza wewnętrznego. Przegląd Elektrotechniczny **92**, 181–184 (2015). https://doi.org/10.15199/48.2016.05.34
22. Li, J., Yang, Z.: A QP-free algorithm without a penalty function or a filter for nonlinear general-constrained optimization. Appl. Math. Comput. **316**, 52–72 (2018). https://doi.org/10.1016/j.amc.2017.08.013
23. Ma, J., Mahapatra, P., Zitney, S.E., Biegler, L.T., Miller, D.C.: D-RM builder: a software tool for generating fast and accurate nonlinear dynamic reduced models from high-fidelity models. Comput. Chem. Eng. **94**, 60–74 (2016). https://doi.org/10.1016/j.compchemeng.2016.07.021
24. Nocedal, J., Wright, S.J.: Numerical Optimization, 2nd edn. Springer, Springer Series in Operation Research and Financial Engineering (2006)
25. Pandelidis, D., Anisimov, S.: Numerical study and optimization of the cross-flow Maisotsenko cycle indirect evaporative air cooler. Int. J. Heat Mass Transf. **103**, 1029–1041 (2016). https://doi.org/10.1016/j.ijheatmasstransfer.2016.08.014
26. Unal, C., Salamci, M.U.: Drug administration in cancer treatment via optimal nonlinear state feedback gain matrix design. IFAC-PapersOnLine **50**(1), 9979–9984 (2017). https://doi.org/10.1016/j.ifacol.2017.08.1594
27. Wan, W., Eason, J.P., Nicholson, B., Biegler, L.T.: Parallel cyclic reduction decomposition for dynamic optimization problems. Comput. Chem. Eng. (2017) (in press). https://doi.org/10.1016/j.compchemeng.2017.09.023
28. Wang, L., Liu, X., Zhang, Z.: An efficient interior-point algorithm with new non-monotone line search filter method for nonlinear constrained programming. Eng. Optim. **49**, 290–310 (2017). https://doi.org/10.1080/0305215X.2016.1176828
29. Xu, P.: Production of chemicals using dynamic control of metabolic fluxes. Curr. Opin. Biotechnol. **53**, 12–19 (2018). https://doi.org/10.1016/j.copbio.2017.10.009

Intercriteria Analysis of ACO Performance for Workforce Planning Problem

Olympia Roeva, Stefka Fidanova, Gabriel Luque and Marcin Paprzycki

Abstract The workforce planning helps organizations to optimize the production process with the aim to minimize the assigning costs. The problem is to select a set of employees from a set of available workers and to assign this staff to the jobs to be performed. A workforce planning problem is very complex and requires special algorithms to be solved. The complexity of this problem does not allow the application of exact methods for instances of realistic size. Therefore, we will apply Ant Colony Optimization (ACO) algorithm, which is a stochastic method for solving combinatorial optimization problems. The ACO algorithm is tested on a set of 20 workforce planning problem instances. The obtained solutions are compared with other methods, as scatter search and genetic algorithm. The results show that ACO algorithm performs better than other the two algorithms. Further, we focus on the influence of the number of ants and the number of iterations on ACO algorithm performance. The tests are done on 16 different problem instances – ten structured and six unstructured problems. The results from ACO optimization procedures are discussed. In order to evaluate the influence of considered ACO parameters additional investigation is done. InterCriteria Analysis is performed on the ACO results for the regarded 16 problems. The results show that for the considered here workforce

O. Roeva
Institute of Biophysics and Biomedical Engineering, Bulgarian Academy of Sciences,
Sofia, Bulgaria
e-mail: olympia@biomed.bas.bg

S. Fidanova (✉)
Institute of Information and Communication Technology,
Bulgarian Academy of Sciences, Sofia, Bulgaria
e-mail: stefka@parallel.bas.bg

G. Luque
Department of Languages and Computer Science, University of Mlaga,
29071 Mlaga, Spain
e-mail: gabriel@lcc.uma.es

M. Paprzycki
System Research Institute,
Polish Academy of Sciences, Warsaw and Management Academy, Warsaw, Poland
e-mail: marcin.paprzycki@ibspan.waw.pl

© Springer Nature Switzerland AG 2019
S. Fidanova (ed.), *Recent Advances in Computational Optimization*,
Studies in Computational Intelligence 795,
https://doi.org/10.1007/978-3-319-99648-6_4

47

planning problem the best performance is achieved by the ACO algorithm with five ants in population.

Keywords Workforce planning · Ant colony optimization · Metaheuristics InterCriteria analysis

1 Introduction

The workforce planing is an essential question of the human resource management. This is an important industrial decision making problem. It is a hard optimization problem, which includes multiple levels of complexity. This problem contains two decision sets: selection and assignment. The first set selects employees from the larger set of available workers. The second set assigns the employees to the jobs to be performed. The aim is minimal assignment cost while the work requirements are fulfilled.

For this very complex problem with strong constraints it is impossible to apply exact methods for instances with realistic size. A deterministic workforce planing problem is studied in [26, 32]. In the work [26] workforce planning models that contain non-linear models of human learning are reformulated as mixed integer programs. The authors show that the mixed integer program is much easier to solve than the non-linear program. In [32] a model of workforce planning, that includes workers differences, as well as the possibility of workers training and improving, is considered. A variant of the problem with random demands is proposed in [19, 33]. In [19] a two-stage stochastic program for scheduling and allocating cross-trained workers is proposed considering a multi-department service environment with random demands. In some problems uncertainty has been employed [27, 30, 31, 39, 41]. In such cases the corresponding objective function and given constraints are converted into crisp equivalents and then the model is solved by traditional methods [31] or the considered uncertain model is transformed into an equivalent deterministic form as it is shown in [39]. Most of the approaches simplify the problem by omitting some of the constraints. Some conventional methods can be applied to workforce planning problem as mixed linear programming [21] and decomposition method [33]. However, for the more complex non-linear workforce planning problems, the convex methods are not applicable. In this case some heuristic methods including genetic algorithm [1, 29], memetic algorithm [37], scatter search [1], are applied.

In this work we propose an Ant Colony Optimization (ACO) algorithm for workforce planning problem. So far the ACO algorithm has been proven to be very effective in solving various complex optimization problems [23, 25]. We consider the variant of the workforce planning problem proposed in [1]. Our ACO algorithm performance is compared with the performance of the genetic algorithm and scatter search shown in [1]. Moreover, we focus on optimization of the algorithm parameters in order to find the minimal number of ants which is needed to find the best solution. It is known that when the number of ant doubles, the computational time and the used memory doubles, too. When the number of iterations doubles, only the computational time doubles. We look for a minimal product between number of ants and number of iterations that is sufficient to find the best solution.

In addition, we apply the recently developed approach – InterCriteria Analysis (ICrA) [11]. ICrA is an approach aiming to go beyond the nature of the criteria involved in a process of evaluation of multiple objects against multiple criteria, and, thus to discover some dependencies between the ICrA criteria themselves [11]. Initially, ICrA has been applied for the purposes of temporal, threshold and trends analyses of an economic case-study of European Union member states' competitiveness [15–17]. Further, ICrA has been used to discover the dependencies of different problems as [38, 40] and analysis of the performance of some metaheuristics as GAs and ACO [2, 22, 35, 36]. Published results show the applicability of the ICrA and the correctness of the approach.

ICrA could be appropriate approach for establishing the correlations between different ACO algorithms, based on their performance. ICrA may lead to additional exploration of the considered here problem. Due to that reason, in this paper, ICrA is applied to facilitate the analysis of the number of ants and number of iterations influence on the ACO performance in the considered here workforce planing problem.

The rest of the paper is organized as follows. In Sect. 2 the mathematical description of the workforce planing problem is presented. In Sect. 3 the ACO algorithm for workforce planing problem is proposed. The theoretical background of the ICrA is given in Sect. 4. The numerical results from ACO application for workforce planing problem are summarized and discussed in Sect. 5. In Sect. 6 ACO performance based on differently tuned algorithm parameters is investigated. The presented results are discussed in terms of which ACO algorithm is best to solve the workforce planing problem. The results from ICrA application are discussed in Sect. 7. In Sect. 8 some conclusions and directions for future works are done.

2 The Workforce Planning Problem

In this paper we use the description of workforce planing problem given by Glover et al. [24]. There is a set of jobs $J = \{1, \ldots, m\}$, which must be completed during a fixed period (week, for example). Every job j requires d_j hours to be completed. The set of available workers is $I = \{1, \ldots, n\}$. For efficiency reason each worker must perform all assigned to him jobs for minimum h_{min} hours. The worker i is available for s_i hours. The maximal number of assigned jobs to the same worker is j_{max}. Workers have different skills and the set A_i shows the jobs that the worker i is qualified to perform. The maximal number of workers which can be assigned during the planed period is t, i.e., at most t workers may be selected from the set I of workers and the selected workers must be capable to complete all the jobs. The aim is to find feasible solution that optimizes the objective function.

Each worker i and job j are related with cost c_{ij} of assigning the worker to the job. The mathematical model of the workforce planing problem is as follows:

$$x_{ij} = \begin{cases} 1 & \text{if the worker } i \text{ is assigned to job } j \\ 0 & \text{otherwise} \end{cases}$$

$$y_i = \begin{cases} 1 & \text{if worker } i \text{ is selected} \\ 0 & \text{otherwise} \end{cases}$$

$$z_{ij} = \text{number of hours that worker } i$$

$$\text{is assigned to perform job } j$$

$$Q_j = \text{set of workers qualified to perform job } j$$

$$\text{Minimize} \sum_{i \in I} \sum_{j \in A_i} c_{ij} \cdot x_{ij} \tag{1}$$

Subject to

$$\sum_{j \in A_i} z_{ij} \leq s_i \cdot y_i \quad i \in I \tag{2}$$

$$\sum_{i \in Q_j} z_{ij} \geq d_j \quad j \in J \tag{3}$$

$$\sum_{j \in A_i} x_{ij} \leq j_{max} \cdot y_j \quad i \in I \tag{4}$$

$$h_{min} \cdot x_{ij} \leq z_{ij} \leq s_i \cdot x_{ij} \quad i \in I, j \in A_i \tag{5}$$

$$\sum_{i \in I} y_i \leq t \tag{6}$$

$$x_{ij} \in \{0, 1\} \ i \in I, j \in A_i$$
$$y_i \in \{0, 1\} \ i \in I$$
$$z_{ij} \geq 0 \qquad i \in I, j \in A_i$$

The objective function of this problem minimizes the total assignment cost. The number of hours for each selected worker is bounded (inequality 2). The work must be done in full (inequality 3). The number of the jobs, that each worker can perform is limited (inequality 4). There is minimal number of hours that each job must be performed by all assigned workers to can work efficiently (inequality 5). The number of assigned workers is limited (inequality 6).

Different objective functions can be optimized with the same model. In this paper our aim is to minimize the total assignment cost. If $\tilde{c}_{ij}$ is the cost the worker i to performs the job j for one hour, than the objective function can minimize the cost of the all jobs to be finished (on hourly basis).

$$f(x) = \text{Min} \sum_{i \in I} \sum_{j \in A_i} \tilde{c}_{ij} . x_{ij} \tag{7}$$

Some workers can have preference to perform part of the jobs he is qualified and the objective function can be to maximize the satisfaction of the workers preferences or to maximize the minimum preference value for the set of selected workers.

As we mentioned above in this paper the assignment cost is minimized (equation 1). This problem is similar to the Capacitate Facility Location Problem (CFLP). The workforce planning problem is difficult to be solved because of very restrictive constraints especially the relation between the parameters h_{min} and d_j. When the problem is structured (d_j is a multiple of h_{min}), it is far easier to find feasible solution, than for unstructured problems (d_j and h_{min} are not related).

3 Ant Colony Optimization

The ACO is a metaheuristic methodology which follows the real ant colonies behaviour when they look for a food and return back to the nest. Real ants use chemical substance, called pheromone, to mark their path ant to be able to return back. An isolated ant moves randomly, but when an ant detects a previously laid pheromone it can decide to follow the trail and to reinforce it with additional quantity of pheromone. The repetition of the above mechanism represents the auto-catalytic behavior of a real ant colony, where the more ants follow a given trail, the more attractive that trail becomes. Thus the ants can collectively find a shorter path between the nest and the source of the food. The main idea of the ACO algorithms comes from this natural behaviour.

3.1 Main ACO Algorithm

Metaheuristic methods are applied on difficult in computational point of view problems, when it is impractical to use traditional numerical methods. A lot of problems coming from real life, especially from the industry. These problems need exponential number of calculations and the only option, when the problem is large, is to apply some metaheuristic methods in order to obtain a good solution for a reasonable time [20].

ACO algorithm is proposed by Dorigo et al. [18]. Later some modifications have been proposed mainly in pheromone updating rules [20]. The artificial ants in ACO algorithms simulate the ants behaviour. The problem is represented by graph. The solutions are represented by paths in a graph and we look for shorter path corresponding to given constraints. The requirements of ACO algorithm are as follows:

– Suitable representation of the problem by a graph;
– Suitable pheromone placement on the nodes or on the arcs of the graph;
– Appropriate problem-dependent heuristic function, which manage the ants to improve solutions;
– Pheromone updating rules;
– Transition probability rule, which specifies how to include new nodes in the partial solution.

The structure of the ACO algorithm is shown on Fig. 1.

The transition probability $p_{i,j}$, to choose the node j, when the current node is i, is a product of the heuristic information $\eta_{i,j}$ and the pheromone trail level $\tau_{i,j}$ related with this move, where $i, j = 1, \ldots, n$.

$$p_{i,j} = \frac{\tau_{i,j}^a \eta_{i,j}^b}{\sum\limits_{k \in Unused} \tau_{i,k}^a \eta_{i,k}^b}, \tag{8}$$

where $Unused$ is the set of unused nodes of the graph.

A node becomes more profitable if the value of the heuristic information and/or the related pheromone is higher. At the beginning, the initial pheromone level is the same for all elements of the graph and is set to a small positive constant value τ_0, $0 < \tau_0 < 1$. At the end of every iteration the ants update the pheromone values. Different ACO algorithms adopt different criteria to update the pheromone level [20].

The main pheromone trail update rule is:

$$\tau_{i,j} \leftarrow \rho\tau_{i,j} + \Delta\tau_{i,j}, \tag{9}$$

where ρ decreases the value of the pheromone, like the evaporation in a nature. $\Delta\tau_{i,j}$ is a newly added pheromone, which is proportional to the quality of the solution. The quality of the solution is measured by the value of the objective function of the solution constructed by the ant.

An ant start to construct its solution from a random node of the graph of the problem. The random start is a diversification of the search. Because of the random start

Fig. 1 Pseudo-code of ACO algorithm

Ant Colony Optimization
Initialize number of ants;
Initialize the ACO parameters;
while not end condition **do**
    for $k = 0$ **to** number of ants
        ant k chooses start node;
        while solution is not constructed **do**
            ant k selects higher probability node;
        end while
    end for
    Update pheromone trails;
end while

a relatively small number of ants can be used, compared to other population based metaheuristics. The heuristic information represents the prior knowledge of the problem, which we use to better manage the ants. The pheromone is a global experience of the ants to find optimal solution. The pheromone is a tool for concentration of the search around the best so far solutions.

3.2 ACO Algorithm for Workforce Planning

One of the essential point of the ant algorithm is the proper representation of the problem by graph. In our case the graph of the problem is 3 dimensional and the node (i, j, z) denotes to worker i to be assigned to the job j for time z. At the beginning of every iteration each ant starts to construct its solution, from random node of the graph of the problem. For each ant are generated three random numbers. The first random number is in the interval $[0, \ldots, n]$ and corresponds to the worker we assign. The second random number is in the interval $[0, \ldots, m]$ and corresponds to the job which this worker will perform. The third random number is in the interval $[h_{min}, \ldots, \min\{d_j, s_i\}]$ and corresponds to the number of hours worker i is assigned to perform the job j. Subsequently, the ant applies the transition probability rule to include next nodes in the partial solution, until the solution is completed.

We propose the following heuristic information:

$$\eta_{ijl} = \begin{cases} 1/c_{ij} & l = z_{ij} \\ 0 & \text{otherwise} \end{cases} \tag{10}$$

This heuristic information stimulates to assign the most cheapest worker as longer as possible. The ant chooses the node with the highest probability. When an ant has several possibilities for next node (several candidates have the same probability to be chosen), the next node is chosen randomly among them.

When a new node is included we take in to account how many workers are assigned currently, how many time slots every worker is currently assigned and how many time slots are currently assigned per job. When some move of the ant does not meet the problem constraints, the probability of this move is set to 0. If it is impossible to include new nodes from the graph of the problem (for all nodes the value of the transition probability is 0), the construction of the solution stops. When the constructed solution is feasible the value of the objective function is the sum of the assignment cost of the assigned workers. If the constructed solution is not feasible, the value of the objective function is set to -1.

Only the ants, which constructed feasible solution are allowed to add new pheromone to the elements of their solutions. The newly added pheromone is equal to the reciprocal value of the objective function:

$$\Delta\tau_{i,j} = \frac{\rho - 1}{f(x)} \tag{11}$$

Thus, the nodes of the problem graph, which belong to better solutions (less value of the objective function) receive more pheromone than the other nodes and become more desirable in the next iteration.

At the end of every iteration we compare the best solution with the best so far solution. If the best solution from the current iteration is better than the best so far solution (global best solution), we update the global best solution with the current iteration best solution.

The end condition used in our ACO algorithm is the number of iterations.

4 InterCriteria Analysis

InterCriteria analysis, based on the apparatuses of index matrices [3, 5, 7–9] and intuitionistic fuzzy sets (IFSs) [4, 6, 10], is given in details in [11]. Here, for completeness, the proposed idea is briefly presented.

An intuitionistic fuzzy pair (IFP) [12] is an ordered pair of real non-negative numbers $\langle a, b \rangle$, where $a, b \in [0, 1]$ and $a + b \leq 1$, that is used as an evaluation of some object or process. According to [12], the components (a and b) of IFP might be interpreted as degrees of "membership" and "non-membership" to a given set, degrees of "agreement" and "disagreement", degrees of "validity" and "non-validity", degrees of "correctness" and "non-correctness", etc.

The apparatus of index matrices is presented initially in [5] and discussed in more details in [7, 8]. For the purposes of ICrA application, the initial index set consists of the criteria (for rows) and objects (for columns) with the index matrix elements assumed to be real numbers. Further, an index matrix with index sets consisting of the criteria (for rows and for columns) with IFP elements determining the degrees of correspondence between the respective criteria is constructed, as it is going to be briefly presented below.

Let the initial index matrix is presented in the form of Eq. (12), where, for every p, q, $(1 \leq p \leq m, 1 \leq q \leq n)$, C_p is a criterion, taking part in the evaluation; O_q – an object to be evaluated; $C_p(O_q)$ – a real number (the value assigned by the p-th criteria to the q-th object).

$$
A = \begin{array}{c|cccccccc}
 & O_1 & \dots & O_k & \dots & O_l & \dots & O_n \\
\hline
C_1 & C_1(O_1) & \dots & C_1(O_k) & \dots & C_1(O_l) & \dots & C_1(O_n) \\
\vdots & \vdots & \ddots & \vdots & \ddots & \vdots & \ddots & \vdots \\
C_i & C_i(O_1) & \dots & C_i(O_k) & \dots & C_i(O_l) & \dots & C_i(O_n) \\
\vdots & \vdots & \ddots & \vdots & \ddots & \vdots & \ddots & \vdots \\
C_j & C_j(O_1) & \dots & C_j(O_k) & \dots & C_j(O_l) & \dots & C_j(O_n) \\
\vdots & \vdots & \ddots & \vdots & \ddots & \vdots & \ddots & \vdots \\
C_m & C_m(O_1) & \dots & C_m(O_k) & \dots & C_m(O_l) & \dots & C_m(O_n)
\end{array}
\qquad (12)
$$

Let O denotes the set of all objects being evaluated, and $C(O)$ is the set of values assigned by a given criteria C (i.e., $C = C_p$ for some fixed p) to the objects, i.e.,

$$O \stackrel{\text{def}}{=} \{O_1, O_2, O_3, \ldots, O_n\},$$

$$C(O) \stackrel{\text{def}}{=} \{C(O_1), C(O_2), C(O_3), \ldots, C(O_n)\}.$$

Let $x_i = C(O_i)$. Then the following set can be defined:

$$C^*(O) \stackrel{\text{def}}{=} \{\langle x_i, x_j \rangle | i \neq j \,\&\, \langle x_i, x_j \rangle \in C(O) \times C(O)\}.$$

Further, if $x = C(O_i)$ and $y = C(O_j)$, $x \prec y$ iff $i < j$ will be written.

In order to find the agreement of different criteria, the vectors of all internal comparisons for each criterion are constructed, which elements fulfil one of the three relations R, $\overline{R}$ and $\tilde{R}$. The nature of the relations is chosen such that for a fixed criterion C and any ordered pair $\langle x, y \rangle \in C^*(O)$:

$$\langle x, y \rangle \in R \Leftrightarrow \langle y, x \rangle \in \overline{R}, \tag{13}$$

$$\langle x, y \rangle \in \tilde{R} \Leftrightarrow \langle x, y \rangle \notin (R \cup \overline{R}), \tag{14}$$

$$R \cup \overline{R} \cup \tilde{R} = C^*(O). \tag{15}$$

For example, if "R" is the relation "$<$", then $\overline{R}$ is the relation "$>$", and vice versa.

Hence, for the effective calculation of the vector of internal comparisons (denoted further by $V(C)$) only the considering of a subset of $C(O) \times C(O)$ is needed, namely:

$$C^{\prec}(O) \stackrel{\text{def}}{=} \{\langle x, y \rangle | x \prec y \,\&\, \langle x, y \rangle \in C(O) \times C(O),$$

due to Eqs. (13)–(15). For brevity, $c^{i,j} = \langle C(O_i), C(O_j) \rangle$.

Then for a fixed criterion C the vector of lexicographically ordered pair elements is constructed:

$$V(C) = \{c^{1,2}, c^{1,3}, \ldots, c^{1,n}, c^{2,3}, c^{2,4}, \ldots, c^{2,n}, c^{3,4}, \ldots, c^{3,n}, \ldots, c^{n-1,n}\}. \tag{16}$$

In order to be more suitable for calculations, $V(C)$ is replaced by $\hat{V}(C)$, where its k-th component ($1 \leq k \leq \frac{n(n-1)}{2}$) is given by:

$$\hat{V}_k(C) = \begin{cases} 1, & \text{iff } V_k(C) \in R, \\ -1, & \text{iff } V_k(C) \in \overline{R}, \\ 0, & \text{otherwise.} \end{cases}$$

When comparing two criteria the degree of "agreement" is determined as the number of matching components of the respective vectors (divided by the length of the vector for normalization purposes). This can be done in several ways, e.g. by counting the matches or by taking the complement of the Hamming distance. The degree of "disagreement" is the number of components of opposing signs in the two vectors (again normalized by the length).

If the respective degrees of "agreement" and "disagreement" are denoted by $\mu_{C,C'}$ and $\nu_{C,C'}$, it is obvious (from the way of computation) that $\mu_{C,C'} = \mu_{C',C}$ and $\nu_{C,C'} = \nu_{C',C}$. Also it is true that $\langle \mu_{C,C'}, \nu_{C,C'} \rangle$ is an IFP.

In the most of the obtained pairs $\langle \mu_{C,C'}, \nu_{C,C'} \rangle$, the sum $\mu_{C,C'} + \nu_{C,C'}$ is equal to 1. However, there may be some pairs, for which this sum is less than 1. The difference

$$\pi_{C,C'} = 1 - \mu_{C,C'} - \nu_{C,C'} \tag{17}$$

is considered as a degree of "uncertainty".

The following index matrix is constructed as a result of applying the ICrA to A [Eq. (12)]:

$$
\begin{array}{c|ccc}
 & C_2 & \cdots & C_m \\
\hline
C_1 & \langle \mu_{C_1,C_2}, \nu_{C_1,C_2} \rangle & \cdots & \langle \mu_{C_1,C_m}, \nu_{C_1,C_m} \rangle \\
\vdots & \vdots & \ddots & \vdots \\
C_{m-1} & & \cdots & \langle \mu_{C_{m-1},C_m}, \nu_{C_{m-1},C_m} \rangle
\end{array},
$$

that determines the degrees of correspondence between criteria $C_1, ..., C_m$.

In this paper we use μ-biased algorithm **Algorithm** 1 for calculation of intercriteria relations [34]. An example pseudocode of the **Algorithm** 1 is presented below.

Algorithm 1 Calculating $\mu_{C,C'}$ and $\nu_{C,C'}$ between two criteria

---

Require: Vectors $\hat{V}(C)$ and $\hat{V}(C')$

1: **function** DEGREES OF AGREEMENT AND DISAGREEMENT($\hat{V}(C), \hat{V}(C')$)
2:    $V \leftarrow \hat{V}(C) - \hat{V}(C')$
3:    $\mu \leftarrow 0$
4:    $\nu \leftarrow 0$
5:    **for** $i \leftarrow 1$ to $\frac{n(n-1)}{2}$ **do**
6:       **if** $V_i = 0$ **then**
7:          $\mu \leftarrow \mu + 1$
8:       **else if** abs(V_i) = 2 **then**   ▷ abs(V_i): the absolute value of V_i
9:          $\nu \leftarrow \nu + 1$
10:       **end if**
11:    **end for**
12:    $\mu \leftarrow \frac{2}{n(n-1)}\mu$
13:    $\nu \leftarrow \frac{2}{n(n-1)}\nu$
14:    **return** μ, ν
15: **end function**

---

5   Application of ACO for Workforce Planning Problem

In this section we report test results and compare them with results achieved by other methods. We analyse the algorithm performance and the quality of the achieved solutions. The software, which realizes the algorithm is written in C and is run on Pentium desktop computer at 2.8 GHz with 4 GB of memory.

We use the artificially generated problem instances considered in [1]. The test instances characteristics are shown in Table 1.

The set of test problems consists of ten structured and ten unstructured problems. The problem is structured when d_j is proportional to h_{min}. The structured problems are enumerated from $S01$ to $S10$ and unstructured problems are enumerated from $U01$ to $U10$.

The number of iterations (stopping criteria) is fixed to be 100. The parameter settings of our ACO algorithm are shown in Table 2. The values are fixed experimentally.

The algorithm is stochastic and from a statistical point of view it needs to be run minimum 30 times to guarantee the robustness of the average results. We perform 30 independent runs of the algorithm. Afterwards statistical analysis of the results applying ANOVA test was done. The test shows that there is significant difference between the results achieved by different methods, or the results are not statistically the same.

Table 1   Test instances characteristics

Parameters	Value
n	20
m	20
t	10
s_i	[50, 70]
j_{max}	[3, 5]
h_{min}	[10, 15]

Table 2   ACO parameter settings

Parameters	Value
Number of iterations	100
ρ	0.5
τ_0	0.5
Number of ants	20
a	1
b	1

Let us compare the numerical results achieved by our ACO algorithm and those achieved by genetic algorithm (GA) and scatter search (SS) presented in [1]. Table 3 shows the achieved results for structured instances, while Table 4 shows the obtained results for unstructured instances.

We observe that ACO algorithm outperforms the other two algorithms. ACO is a constructive method and when the graph of the problem and heuristic information are appropriate and they represent the problem in a good way, it can help a lot of for better algorithm performance and achieving good solutions. Our graph of the problem has a star shape. Each worker and job are linked with several nodes, corresponding to the time, for which the worker is assigned to perform this job. The proposed heuristic information stimulates the cheapest workers to be assigned for longer time. It is a greedy strategy. After the first iteration the pheromone level reflects the experience

Table 3 Average results for structured problems

Test problem	Objective function value		
	SS	GA	ACO
S01	936	963	807
S02	952	994	818
S03	1095	1152	882
S04	1043	1201	849
S05	1099	1098	940
S06	1076	1193	869
S07	987	1086	812
S08	1293	1287	872
S09	1086	1107	793
S10	945	1086	825

Table 4 Average results for unstructured problems

Test problem	Objective function value		
	SS	GA	ACO
U01	1586	1631	814
U02	1276	1264	845
U03	1502	1539	906
U04	1653	1603	869
U05	1287	1356	851
U06	1193	1205	873
U07	1328	1301	828
U08	1141	1106	801
U09	1055	1173	768
U10	1178	1214	818

of the ants during the searching process thus affects the strategy. The elements of good solutions accumulate more pheromone, during the algorithm performance, than others and become more desirable in the next iterations.

Now we will compare the execution time of the proposed ACO algorithm with the execution time of the other two algorithms – GA and SS [1]. The algorithms are run on similar computers. In Table 5 the parameters of the GA and SS algorithms are presented.

The average execution times over 30 runs of each of the algorithms are reported in Tables 6 and 7.

It is seen that the ACO algorithm finds the solution faster than GA and SS. Considering the execution time the GA and SS algorithms have similar performance. By the numerical results presented in Tables 3, 4, 6 and 7 we can conclude that ACO algorithm gives very encouraging results. It achieves better solutions in shorter time than

Table 5 Algorithms parameter settings, as given in [1]

Parameters	Genetic algorithm
Population size	400
Crossover rate	0.8
Mutation rate	0.2
Parameters	Scatter search
Initial population	15
Reference set update and creation	8
Subset generated	All 2-elements subsets
ρ_i	0.1

Table 6 Average time for structured problems

Test problem	Execution time, s		
	SS	GA	ACO
$S01$	72	61	26
$S02$	49	32	21
$S03$	114	111	22
$S04$	86	87	25
$S05$	43	40	21
$S06$	121	110	23
$S07$	52	49	23
$S08$	46	42	24
$S09$	70	67	20
$S10$	105	102	22

Table 7 Average time for unstructured problems

Test problem	Execution time, s		
	SS	GA	ACO
$U01$	102	95	22
$U02$	94	87	20
$U03$	58	51	20
$U04$	83	79	20
$U05$	62	57	23
$U06$	111	75	22
$U07$	80	79	21
$U08$	123	89	20
$U09$	75	72	26
$U10$	99	95	20

the other two algorithms, SS and GA. If we compare the used memory, the ACO algorithm uses less memory than GA (GA population size is 400 individuals) and similar memory to SS (initial population size is 15 and reference set is 8 individuals) [1].

6 Influence of ACO Parameters on Algorithm Performance

In this section we analyse the ACO performance according to the number of ants and the quality of the achieved solutions. We use the same artificially generated problem instances, considered in [1].

If the number of ants of ACO algorithm increases, the computational time and the used memory increase proportionally. If the number of iterations increases, only the computational time increases. If the computational time is fixed and we vary only the number of ants it means that we vary the number of iteration too, but in opposite direction, or if the time is fixed it is equivalent to fixing the product of number of ants and number of iterations.

We apply number of ants from the set {5, 10, 20, 40} and respectively, number of iteration – {400, 200, 100, 50}. Because of stochastic nature of the algorithm we run the ACO algorithm for all 16 test problems with each one of the four ACO algorithms ($ACO_{5\times400}$, $ACO_{10\times200}$, $ACO_{20\times100}$ and $ACO_{40\times50}$) 30 times. We look for the maximal number of iterations, within the fixed computational time, which is needed to find the best solution. We compare the product between the number of ants and number of required iterations for the best performed ACO algorithm. The parameter settings for our ACO algorithm are shown in Table 8.

Tables 9 and 10 show the resulting product between the number of ants and number of iterations that have been used to find the best solution. When the observed product is the same for different ACO algorithms (with different number of ants) the best

Table 8 ACO parameter settings

Parameters	Value
Number of iterations	400, 200, 100, 50
ρ	0.5
τ_0	0.5
Number of ants	5, 10, 20, 40
a	1
b	1

Table 9 Computational results for structured problems

Test problem	Product ants × Iterations			
	$ACO_{5 \times 400}$	$ACO_{10 \times 200}$	$ACO_{20 \times 100}$	$ACO_{40 \times 50}$
$S01$	195	200	260	**120**
$S02$	**195**	330	340	640
$S03$	**475**	490	580	1160
$S04$	**1540**	1540	1540	1560
$S05$	415	**320**	420	920
$S06$	**165**	250	420	520
$S07$	**570**	720	860	880
$S08$	**1125**	1130	1140	1120
$S09$	**855**	860	720	880
$S10$	**230**	230	520	840

Table 10 Computational results for unstructured problems

Test problem	Product ants × Iterations			
	$ACO_{5 \times 400}$	$ACO_{10 \times 200}$	$ACO_{20 \times 100}$	$ACO_{40 \times 50}$
$U00$	**775**	780	780	2060
$U01$	**330**	340	340	440
$U02$	**160**	340	340	400
$U03$	1000	1000	**560**	640
$U04$	**760**	760	820	820
$U05$	295	295	**280**	280

results is this produced by ACO with lesser number of ants, because in this case the used memory is less. The best results are shown in bold.

Let us discuss the results reported in Tables 9 and 10. Regarding the structured problems, for 8 of them the best execution time is when the number of ants is 5. Only for two of the test problems ($S01$ and $S05$) the execution time is better for 40 and 10 ants, respectively, but the result is close to this achieved with 5 ants. Regarding unstructured test problems, for four of them the best execution time is achieved when the number of used ants is 5. For two test problems ($U03$ and $U05$) the best results are achieved using ACO with 20 ants. For test $U05$, the result is quite close to the result with 5 ants. Only for test $U03$ the difference with the results using 5 ants is significant. Thus, we can conclude that for this problem the best algorithm performance, using less computational resources, is when the number of used ants is 5.

7 InterCriteria Analysis of the Results

In this section we use ICrA to obtain some additional knowledge about considered four ACO algorithms. Based on the results in Tables 9 and 10 we construct the following index matrix, Eq. (18):

	$ACO_{5 \times 400}$	$ACO_{10 \times 200}$	$ACO_{20 \times 100}$	$ACO_{40 \times 50}$
S01	195	200	260	120
S02	195	330	340	640
S03	475	490	580	1160
S04	1540	1540	1540	1560
S05	415	320	420	920
S06	165	250	420	520
S07	570	720	860	880
S08	1125	1130	1140	1120
S09	855	860	720	880
S10	230	230	520	840
U00	775	780	780	2060
U01	330	340	340	440
U02	160	340	340	400
U03	1000	1000	560	640
U04	760	760	820	820
U05	295	295	280	280

$$(18)$$

From Eq. (18) it can be seen that the ICrA objects (S01, S02, ..., S10, U00, ..., U05) are the different test problems and the ICrA criteria ($ACO_{5 \times 400}$, $ACO_{10 \times 200}$, $ACO_{20 \times 100}$ and $ACO_{40 \times 50}$) are the ACO algorithms with different number of ants.

After application of ICrA, using the software ICrAData [28], to index matrix Eq. (18) we obtained the two index matrices with the relations between considered four criteria. The resulting index matrices for $\mu_{C,C'}$, $\nu_{C,C'}$ and $\pi_{C,C'}$ values are shown in Eqs. (19)–(21).

$\mu_{C,C'}$	$ACO_{5\times400}$	$ACO_{10\times200}$	$ACO_{20\times100}$	$ACO_{40\times50}$	
$ACO_{5\times400}$	1	0.88	0.78	0.74	
$ACO_{10\times200}$	0.88	1	0.78	0.71	(19)
$ACO_{20\times100}$	0.78	0.78	1	0.81	
$ACO_{40\times50}$	0.74	0.71	0.81	1	

$\nu_{C,C'}$	$ACO_{5\times400}$	$ACO_{10\times200}$	$ACO_{20\times100}$	$ACO_{40\times50}$	
$ACO_{5\times400}$	0	0.10	0.18	0.23	
$ACO_{10\times200}$	0.10	0	0.20	0.27	(20)
$ACO_{20\times100}$	0.18	0.20	0	0.14	
$ACO_{40\times50}$	0.23	0.27	0.14	0	

$\pi_{C,C'}$	$ACO_{5\times400}$	$ACO_{10\times200}$	$ACO_{20\times100}$	$ACO_{40\times50}$	
$ACO_{5\times400}$	0	0.02	0.04	0.03	
$ACO_{10\times200}$	0.02	0	0.03	0.03	(21)
$ACO_{20\times100}$	0.04	0.03	0	0.05	
$ACO_{40\times50}$	0.03	0.03	0.05	0	

The obtained ICrA results are visualized on Fig. 2 within the specific triangular geometrical interpretation of IFSs.

The results show that the following criteria pairs, according to [13], are in:

- positive consonance:
  $ACO_{5\times400}$-$ACO_{10\times200}$ – with degree of "agreement"
  $\mu_{C,C'} = 0.88$;
- weak positive consonance:
  $ACO_{20\times100}$-$ACO_{40\times50}$,
  $ACO_{20\times100}$-$ACO_{5\times400}$ and

Fig. 2 Presentation of ICrA results in the intuitionistic fuzzy interpretation triangle

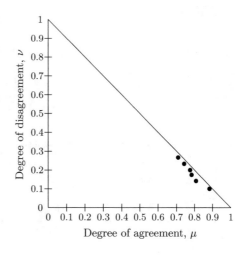

$ACO_{20 \times 100}$-$ACO_{10 \times 200}$ – with degree of "agreement"
$\mu_{C,C'} = 0.81$, $\mu_{C,C'} = 0.78$ and $\mu_{C,C'} = 0.78$, respectively;
– weak dissonance:
$ACO_{40 \times 50}$-$ACO_{5 \times 400}$ and
$ACO_{40 \times 50}$-$ACO_{10 \times 200}$ – with degree of "agreement"
$\mu_{C,C'} = 0.74$ and $\mu_{C,C'} = 0.71$, respectively.

ACO algorithms with close values of number of ants (5 and 10, 10 and 20, 20 and 40) show similar performance. The same result is obtained for ACO algorithms with 5 and 20 ants. The ACO algorithms with bigger difference in the number of ants (5 and 40, 10 and 40) are in weak dissonance, i.e. their performance is not similar (Table 11).

In [14] the author propose to rank the criteria pairs in both dimensions simultaneously (degrees of "agreement" $\mu_{C,C'}$ and "disagreement" $\nu_{C,C'}$ of the intuitionistic fuzzy pairs). This can be done by calculation for each point in the Fig. 2 its distance from the point $\langle 1, 0 \rangle$. The formula for the distance $d_{C,C'}$ of the pair C, C' to the $\langle 1, 0 \rangle$ point is:

$$d_{C,C'} = \sqrt{(1 - \mu_{C,C'})^2 + \nu_{C,C'}{}^2} \qquad (22)$$

The results are presented in Table 12.

Table 11 Consonance and dissonance scale, according to [13]

Interval of $\mu_{C,C'}$	Meaning
(0.00–0.05]	Strong negative consonance
(0.05–0.15]	Negative consonance
(0.15–0.25]	Weak negative consonance
(0.25–0.33]	Weak dissonance
(0.33–0.43]	Dissonance
(0.43–0.57]	Strong dissonance
(0.57–0.67]	Dissonance
(0.67–0.75]	Weak dissonance
(0.75–0.85]	Weak positive consonance
(0.85–0.95]	Positive consonance
(0.95–1.00]	Strong positive consonance

Table 12 Index matrix of criteria distances from point $\langle 1, 0 \rangle$

	$ACO_{5 \times 400}$	$ACO_{10 \times 200}$	$ACO_{20 \times 100}$	$ACO_{40 \times 50}$
$ACO_{5 \times 400}$	0	0.154	0.279	0.348
$ACO_{10 \times 200}$	0.154	0	0.301	0.395
$ACO_{20 \times 100}$	0.279	0.301	0	0.238
$ACO_{40 \times 50}$	0.348	0.395	0.238	0

In this case the criteria pairs (different ACO algorithms) are ordered according to their $d_{C,C'}$ sorted in decreasing order, as follows:

- $ACO_{5\times400}$-$ACO_{10\times200}$;
- $ACO_{20\times100}$-$ACO_{40\times50}$;
- $ACO_{5\times400}$-$ACO_{20\times100}$;
- $ACO_{10\times200}$-$ACO_{20\times100}$;
- $ACO_{5\times400}$-$ACO_{40\times50}$;
- $ACO_{40\times50}$-$ACO_{10\times200}$.

As it can be seen the similar performance of ACO algorithms with 5 and 10, and 20 and 40 ants is approved by these results too. The similarity between ACO with 5 and ACO with 20 ants is observed again. The next three criteria pairs, are ranked in the same manner, too. Thus the obtained ICrA results are confirmed by two different approaches – when using the scale, proposed in [13] and according simultaneously to the degrees of "agreement"$\mu_{C,C'}$ and "disagreement" $\nu_{C,C'}$ of the intuitionistic fuzzy pairs [14].

On the other hand, ICrA confirms the conclusion that for this problem the best algorithm performance, i.e., using less computational resources, is shown ACO algorithm with five number of ants.

8 Conclusion

In this article we propose ACO algorithm for solving workforce planning problem. We compare the performance of our algorithm with other two metahuristic methods, genetic algorithm and scatter search. The comparison is done by various criteria. We observed that ACO algorithm achieves better solutions than the two other algorithms. Regarding the execution time the ACO algorithm is faster. The ACO population consists of 20 individuals and the memory used by the algorithm is similar to the one used by the SS and less than the memory used by the GA. We achieved very encouraging results. As a future work we will combine our ACO algorithm with appropriate local search procedure for eventual further improvement of the algorithm performance and solutions quality.

In this paper we solve workforce planning problem applying ACO algorithm. We focus on influence of the algorithm parameters on its performance. We try to find minimal number of ants and minimal execution time and memory, which are needed to find best solution. The results show that for most of the test problems, the minimal computational resources are used when the number of ants is five. As a future work we will combine our ACO algorithm with appropriate local search procedure for eventual further improvement of the algorithm performance and solutions quality.

Acknowledgements Work presented here is partially supported by the National Scientific Fund of Bulgaria under grants DFNI-DN 02/10 "New Instruments for Knowledge Discovery from Data, and their Modelling" and DFNI-DN 12/5 "Efficient Stochastic Methods and Algorithms for Large Scale Problems".

References

1. Alba, E., Luque, G., Luna, F.: Parallel metaheuristics for workforce planning. J. Math. Model. Algorithms **6**(3), 509–528 (2007)
2. Angelova, M., Roeva, O., Pencheva, T.: InterCriteria analysis of crossover and mutation rates relations in simple genetic algorithm. In: Proceedings of the 2015 Federated Conference on Computer Science and Information Systems, vol. 5, pp. 419–424 (2015)
3. Atanassov, K.: Index Matrices: Towards an Augmented Matrix Calculus. Springer, Switzerland (2014)
4. Atanassov, K.: Intuitionistic fuzzy sets. VII ITKR session, Sofia, 20–23 June 1983. Int. J. Bioautom. **20**(S1), S1–S6 (2016)
5. Atanassov, K.: Generalized index matrices. Comptes rendus de l'Academie bulgare des Sciences **40**(11), 15–18 (1987)
6. Atanassov, K.: On Intuitionistic Fuzzy Sets Theory. Springer, Berlin (2012)
7. Atanassov, K.: On index matrices, Part 1: standard cases. Adv. Stud. Contemp. Math. **20**(2), 291–302 (2010)
8. Atanassov, K.: On index matrices, Part 2: intuitionistic fuzzy case. Proc. Jangjeon Math. Soc. **13**(2), 121–126 (2010)
9. Atanassov, K.: On index matrices. Part 5: 3-dimensional index matrices. Adv. Stud. Contemp. Math. **24**(4), 423–432 (2014)
10. Atanassov, K.: Review and new results on intuitionistic fuzzy sets, mathematical foundations of artificial intelligence seminar, Sofia, 1988, Preprint IM-MFAIS-1-88. Int. J. Bioautom. **20**(S1), S7–S16 (2016)
11. Atanassov, K., Mavrov, D., Atanassova, V.: Intercriteria decision making: a new approach for multicriteria decision making, based on index matrices and intuitionistic fuzzy sets. Issues in on Intuitionistic Fuzzy Sets and Generalized Nets **11**, 1–8 (2014)
12. Atanassov, K., Szmidt, E., Kacprzyk, J.: On intuitionistic fuzzy pairs. Notes Intuitionistic Fuzzy Sets **19**(3), 1–13 (2013)
13. Atanassov, K., Atanassova, V., Gluhchev, G.: InterCriteria analysis: ideas and problems. Notes on Intuitionistic Fuzzy Sets **21**(1), 81–88 (2015)
14. Atanassova, V.: Interpretation in the intuitionistic fuzzy triangle of the results, obtained by the InterCriteria analysis. In: Proceedings of the 9th Conference of the European Society for Fuzzy Logic and Technology (EUSFLAT), pp. 1369–1374 (2015)
15. Atanassova, V., Mavrov, D., Doukovska, L., Atanassov, K.: Discussion on the threshold values in the InterCriteria decision making approach. Notes on Intuitionistic Fuzzy Sets **20**(2), 94–99 (2014)
16. Atanassova, V., Doukovska, L., Atanassov, K., Mavrov, D.: Intercriteria decision making approach to EU member states competitiveness analysis. In: Proceedings of the International Symposium on Business Modeling and Software Design - BMSD'14, pp. 289–294 (2014)
17. Atanassova, V., Doukovska, L., Karastoyanov, D., Capkovic, F.: InterCriteria decision making approach to EU member states competitiveness analysis: trend analysis. In: Intelligent Systems'2014, Advances in Intelligent Systems and Computing, vol. 322, pp. 107–115 (2014)
18. Bonabeau, E., Dorigo, M., Theraulaz, G.: Swarm Intelligence: From Natural to Artificial Systems. Oxford University Press, New York (1999)
19. Campbell, G.: A two-stage stochastic program for scheduling and allocating cross-trained workers. J. Oper. Res. Soc. **62**(6), 1038–1047 (2011)
20. Dorigo, M., Stutzle, T.: Ant Colony Optimization. MIT Press, Cambridge (2004)
21. Easton, F.: Service completion estimates for cross-trained workforce schedules under uncertain attendance and demand. Prod. Oper. Manage. **23**(4), 660–675 (2014)
22. Fidanova, S., Roeva, O., Paprzycki, M.: InterCriteria analysis of ACO start strategies. In: Proceedings of the 2016 Federated Conference on Computer Science and Information Systems, vol. 8, pp. 547–550 (2016)

23. Fidanova, S., Roeva, O., Paprzycki, M., Gepner, P.: InterCriteria analysis of ACO start startegies. In: Proceedings of the 2016 Federated Conference on Computer Science and Information Systems, pp. 547–550 (2016)
24. Glover, F., Kochenberger, G., Laguna, M., Wubbena, T.: Selection and assignment of a skilled workforce to meet job requirements in a fixed planning period. In: MAEB04, pp. 636–641 (2004)
25. Grzybowska, K., Kovcs, G.: Sustainable supply chain—Supporting tools. In: Proceedings of the 2014 Federated Conference on Computer Science and Information Systems, vol. 2, pp. 1321–1329 (2014)
26. Hewitt, M., Chacosky, A., Grasman, S., Thomas, B.: Integer programming techniques for solving non-linear workforce planning models with learning. Eur. J. Oper. Res. **242**(3), 942–950 (2015)
27. Hu, K., Zhang, X., Gen, M., Jo, J.: A new model for single machine scheduling with uncertain processing time. J. Intell. Manufact. **28**(3), 717–725 (2015)
28. Ikonomov, N., Vassilev, P., Roeva, O.: ICrAData software for InterCriteria analysis. Int. J. Bioautom. **22**(2) (2018) (in press)
29. Li, G., Jiang, H., He, T.: A genetic algorithm-based decomposition approach to solve an integrated equipment-workforce-service planning problem. Omega **50**, 1–17 (2015)
30. Li, R., Liu, G.: An uncertain goal programming model for machine scheduling problem. J. Intell. Manufact. **28**(3), 689–694 (2014)
31. Ning, Y., Liu, J., Yan, L.: Uncertain aggregate production planning. Soft Comput. **17**(4), 617–624 (2013)
32. Othman, M., Bhuiyan, N., Gouw, G.: Integrating workers' differences into workforce planning. Comput. Ind. Eng. **63**(4), 1096–1106 (2012)
33. Parisio, A., Jones, C.N.: A two-stage stochastic programming approach to employee scheduling in retail outlets with uncertain demand. Omega **53**, 97–103 (2015)
34. Roeva, O., Vassilev, P., Angelova, M., Su, J., Pencheva, T.: Comparison of different algorithms for InterCriteria relations calculation. In: 2016 IEEE 8th International Conference on Intelligent Systems, pp. 567–572 (2016)
35. Roeva, O., Fidanova, S., Paprzycki, M.: InterCriteria analysis of ACO and GA hybrid algorithms. Stud. Comput. Intell. **610**, 107–126 (2016)
36. Roeva, O., Fidanova, S., Vassilev, P., Gepner, P.: InterCriteria analysis of a model parameters identification using genetic algorithm. Proceedings of the Federated Conference on Computer Science and Information Systems **5**, 501–506 (2015)
37. Soukour, A., Devendeville, L., Lucet, C., Moukrim, A.: A Memetic algorithm for staff scheduling problem in airport security service. Expert Syst. Appl. **40**(18), 7504–7512 (2013)
38. Todinova, S., Mavrov, D., Krumova, S., Marinov, P., Atanassova, V., Atanassov, K., Taneva, S.G.: Blood plasma thermograms dataset analysis by means of InterCriteria and correlation analyses for the case of colorectal cancer. Int. J. Bioautom. **20**(1), 115–124 (2016)
39. Yang, G., Tang, W., Zhao, R.: An uncertain workforce planning problem with job satisfaction. Int. J. Mach. Learn. Cybern. (2016). https://doi.org/10.1007/s13042-016-0539-6
40. Zaharieva, B., Doukovska, L., Ribagin, S., Radeva, I.: InterCriteria decision making approach for Behterev's disease analysis. Int. J. Bioautom. **22**(2) (2018) (in press)
41. Zhou, C., Tang, W., Zhao, R.: An uncertain search model for recruitment problem with enterprise performance. J. Intell. Manufact. **28**(3), 295–704 (2014)

Is Prüfer Code Encoding Always a Bad Idea?

H. Hildmann, D. Y. Atia, D. Ruta and A. F. Isakovic

Abstract Real world problems are often of a complexity that renders deterministic approaches intractable. In the area of applied optimization, heuristics can offer a viable alternative. While potentially forfeiting on finding the most *optimal* solution, these techniques return *good* solutions in a short time. To do so, a suitable modelling of the problem as well as an efficient mapping of the problem's solutions into a so-called *solution space* is required. Since it is very common to represent solutions as graphs, algorithms that efficiently map graphs into a heuristic-friendly solutions-space are of general interest to community. For a special type of graph, namely *trees* (i.e., undirected, connected and acyclic graphs such as Cayley in Phil Mag 13:172–176 (1857) [13]) *Prüfer Code* (Prüfer in Archiv der Mathematik und Physik 27:742–744 (1918) [43]) (PC) offers a bijective encoding process that comes at a low complexity (algorithms of $\Theta(n)$-complexity are known Micikevičius et al. in Linear-time algorithms for encoding trees as sequences of node labels (2007) [37]) and facilitates mapping to $n - 2$ dimensional Euclidean space. However, this encoding does not preserve properties such as e.g., locality and has therefore been shown to be sub-optimal for entire classes of problems (Gottlieb et al. in Prüfer numbers: a poor representation of spanning trees for evolutionary search (2001) [24]). We argue that Prüfer Code does preserve some characterizing properties (e.g., degree of branching and

H. Hildmann (✉)
Dep. de Ingeniería de Sistemas y Automática,
Universidad Carlos III de Madrid (UC3M), Av. Universidad 30,
28911 Léganes, Spain
e-mail: hanno.hildmann@uc3m.es

D. Y. Atia
Khalifa University of Science and Technology, 127788 Abu Dhabi, UAE
e-mail: dina.atia@ku.ac.ae

D. Ruta
EBTIC, Khalifa University of Science and Technology, 127788 Abu Dhabi, UAE
e-mail: dymitr.ruta@ku.ac.ae

A. F. Isakovic
Physics Department, Khalifa University of Science and Technology,
127788 Abu Dhabi, UAE
e-mail: iregx137@gmail.com; abdel.isakovic@ku.ac.ae

© Springer Nature Switzerland AG 2019
S. Fidanova (ed.), *Recent Advances in Computational Optimization*,
Studies in Computational Intelligence 795,
https://doi.org/10.1007/978-3-319-99648-6_5

branching vertices) and that these are sufficiently relevant for certain types of problems to motivate encoding them in PC. We present our investigations and provide an example problem where PC encoding worked very well.

1 Introduction and Outline

The word *heuristic* comes from the Greek $\varepsilon\upsilon\sigma\acute{\iota}\rho\kappa\omega$: "*to find*", "*to discover*"). It describes approaches that *find* or *estimate* good solutions to problems, as opposed to reliably determining the best one. For complex problems it is often impossible to exhaustively check all possible solutions, motivating the use of a heuristic. Furthermore, many problems require only a certain quality of the solution, and investing resources in improving a solution past this point does not add any benefit.

If a heuristic exploits a property common to many problems then it is called a *meta-heuristic* because it can be applied to an entire class of problems. It is not uncommon for meta-heuristics to be inspired by naturally occurring phenomena, which are often taken from the fields of physics and biology. The authors' past work has often been inspired by nature, specifically by social insects. **The reader will find a short overview over heuristics and our favorite approaches in** Sect. 2.

Heuristics rely on some underlying ordering inherent to the solution space to navigate it. They identify acceptable solutions and then continuously and *iteratively* try to improve on them. In order to be able to *move* from one solution to a better one, there has to be some relation between them, which the heuristic can exploit.

Consider, for example, the problem of finding the closest house to a location. We know that house numbers are ordered by the "$<$" relationship, and let's assume that for any *solution* (any member of the natural numbers between 1 and the last house number in the street) we can check how far it is from the location. We therefore can use the fact that neighbouring solutions (in the solution space $\mathbb{N}^+$) have similar performance, because the distance to the location between two neighbours is very similar). It is easy to see that one would not have to look at all houses in the street to find the closest one to a location. Note that, although seemingly trivial, we performed a crucial step here when we *mapped* the individual solutions to the domain of the natural numbers (instead of e.g., the name of the house owner). *Modelling* a problem and *encoding* its solutions (i.e., the mapping into a domain) are very important decisions in the process. There are many ways to represent solutions and we will only focus on one: graphs, and in our case, simple, undirected, connected and acyclic graphs, commonly called *trees* [13]. **In** Sect. 3 **we provide some background on trees and discuss known complexity results** as well as a specific encoding that allows us to represent trees as unique sequences of numbers: *Prüfer Code* [43] (PC).

Unfortunately, there is evidence from the literature that when using PC to encode the solution space of a problem, certain properties (such as e.g., locality [24]) are not preserved. Since these properties may be relevant, PC is considered a sub-optimal encoding for heuristics. **In** Sect. 4 **we impose structure on Prüfer Code (PC) and identify properties that can – in fact – be preserved.** We argue that for entire classes of problems, these properties are actually sufficient to motivate the encoding of solutions in PC. **We support this claim in** Sect. 5 **by referencing to an implementation where PC was used, successfully**. The performance of our implementation was not possible if important properties were not preserved by our encoding. This practical consideration puts the results presented in [24] in perspective.

2 Nature-Inspired Heuristics

Nature inspired computing has long since applied principles and approaches found in nature to increase the accuracy of models and to improve algorithms (cf. [40]) and there is evidence of nature-inspired approaches (e.g., [5]) being extremely efficient [9]. Especially the behaviour of social insects such as ants, termites and bees has been studied extensively and the underlying principles have been applied to a variety of problems (e.g., client-server allocation [29] or resource allocation [28]). Collective behaviour (e.g., clustering, collaboration and signal processing) based on self-organization [12] has been shown in group-living animals from insects to vertebrates and even at the cell level [39]. These findings have stimulated engineers to investigate approaches for the coordination of (semi-)autonomous multi-agent systems (e.g., [1]) based on self-organization [25]. The social insect metaphor for problem solving emphasizes distributedness, interaction between members of a population, flexibility and robustness [30]. The growing understanding of complex collective behaviour offers the prospect of creating artificial systems which are controlled by emergent collective behaviour [49].

The distributed nature of self-organization is one of its defining characteristics and may be its most advantageous feature. Distributed sensing can be achieved using only rudimentary cognition may be widespread across biological taxa, in addition to being appropriate and cost-effective for robotic agents [6]. Distributed methods are also used to determine e.g., paths for agents in large populations. Global goals, such as the social optimum, can be achieved by local processing using only local information [34]. Furthermore, behavioural features can induce positive feedback, potentially leading to collective decision-making in a social context [47].

Reliable information is a crucial factor influencing decision-making. One characteristic that almost certainly has contributed to insect societies' great ecological success is the ability of individuals to work together [2]. The ability to communicate information to other members of the population is considered a key factor. The field of swarm intelligence, which often embraces the idea of fragmented knowledge [8] (information being kept by the population but not necessarily by all individuals), offers an alternative way of designing intelligent systems in which autonomy, emergence and distributed functioning replace control, pre-programming and centralization. Individual members of a swarm typically use only local (onboard) sensing and coordinate their activity via the shared environment [53].

However, explicitly communicated information is not the only information available: in nature, a common source of information comes from inadvertent cues and information produced by the behaviour of peers. When investigated under the the right set-up, communication strategies for members of large robot populations have been shown to evolve to regulate information provided by such cues [38] and it has been suggested that observational learning (the acquisition of information and behaviours through the observation of the behaviour of peers) could be applied to swarms of robots [50]. For example, recent work [48] investigates the impact of minor behavioural individuality on swarms of construction robots.

2.1 Heuristics and Meta-Heuristics

Very generically speaking, heuristics [42] are techniques to derive a small set of candidate solutions (potential improvements) from a given one (normally taken from a set of known good solutions found in previous iterations). This is achieved by imposing some ordering on the elements in the solutions space and by using knowledge to determine in which direction (in the ordering) to focus the search. If stated abstractly enough to be applied to entire classes of problems, these techniques are referred to as *meta*-heuristics. The field of meta-heuristics is broad and complex and the scope of this chapter does not allow an overview to do it justice. Owing to this, we will simply mention some of the authors' favorites and invite the reader to read up on this fascinating area. For a useful compilation of nature-inspired approaches (including the algorithms) we suggest [11].

2.1.1 Ant-Colony Optimization (ACO) [9, 18]

Ants in nature secrete a pheromone when they travel and thereby share information about their paths using their environment as a collective local memory. Very simply speaking: in the physical world we inhabit, the shortest path gets you to your destination faster, therefore the path most frequently travelled (while reaching the destination) is the shortest. ACO has been successfully applied to e.g., adaptive routing in networks [19, 44] or to solve traffic signal control problems [46].

2.1.2 Particle Swarm Optimization (PSO) [32, 51]

Particle Swarm Optimization (PSO) is a dynamic algorithm where particles embodying the candidate solutions can move through the solution space along various directions and with different velocities, thereby imitating various natural swarms of birds, insects or fish [36, 54]. Based on the underlying assumption that good solutions are clustered, enabling the particles to converge (like a flock of birds) on one area in (solution-)space focuses the search effort in regions where good solutions are more likely to be found. This technique has been used to guide groups of computational agents optimizing e.g., wireless sensor networks [33], load balancing or minimizing the cost of heating in the context of system planning [35].

2.1.3   Genetic Algorithms (GA) [20, 21, 52]

Applying the idea of mutation and cross-over combination (as found in biological off-spring generation) to strings representing solutions of computational problems is the equivalent of generating new solutions using parts of previously found good solutions. This has yielded good results for e.g., optimal classification model design, especially in the context of the classifier fusion [22, 36].

2.2   Practical Decisions for the Use of Heuristics

Two aspects can be separated in the application of such an optimization technique:

(1) mapping the problem (and the corresponding domain of solutions) to a so-called *solution-space* (the collection of all solutions), and
(2) the identification of the underlying principle that enables a system (e.g., a collective of agents (ACO) or a single process) to navigate through a solution-space in a way that drives the search towards increasingly good solutions.

The mapping has to be computationally cheap and strip away all non-relevant aspects while keeping those correlations that matter preserved; the used principle has to be able to navigate through the solution-space using these correlations to *sniff out* good solutions. For example, an adequate solution-space for the problem of finding my friend's house number is $\mathbb{N}$, with each member of the set corresponding uniquely to a house number. The correlation between the solution space and the real world is that consecutive even (or odd) numbers are *neighbours* in the real world, i.e., that proximity in the search-space corresponds to some measure of closeness in the real world. If we have the ability to calculate the distance (in meters) from any given house number to your friend's house, then we can leap-frog through $\mathbb{N}$ using the feedback from the occasional check of how far we are away from our destination.

The choice for the representation of solutions is crucial. Heuristics are iterative processes that are guided by previous performances. Therefore, the ability to translate a candidate solution from its representation in the solution space to the form in which its performance can be evaluated, and to do so quickly and efficiently, has a large impact on the speed – and thus the performance – of the system. If this translation incurs long delays or requires a lot of computational power, then assigning a performance value to a candidate solution will be a costly exercise.

Heuristics rely on *trial and error* in the sense that a number of solutions are considered, translated for evaluation and then this set is culled to retain only the better members. This means in each round, a number of solutions will incur costs without being retained in the surviving set. Therefore, the cost of translating and evaluating a solution directly impacts the feasible size of the candidate set. Since this size is often a relevant tuning parameter efficient translations are needed.

However, the fastest translation algorithm is useless when there is no known relation between the solutions in the solution space: if for all intents and purposes the good solutions are distributed randomly throughout the solution space then no guided search can be performed. This means that solutions that share positive features have to end up near other solutions sharing this feature (note: this refers to the corresponding inverse translation, translating solutions *into* the solution space). Therefore, that translation has to be *property preserving*, at least with respect to those properties that impact a solution's performance.

3   Graphs, Trees and Prüfer Code

3.1   Graphs and Trees

A *graph G* is a pair $G = (V, E)$ of two sets: the set $V = \{v_1, \ldots, v_n\}$ of *n vertices* (which are also often referred to as *nodes* or *worlds*) and the set $E = \{e_1, \ldots, e_m\}$ of *m edges* (often called *lines* or *connections*). Each edge e_i is a tuple of two vertices, representing the two vertices that this edge connects (cf. [7, 17]).

One sub-category of graphs are *connected graph without cycles* (i.e., the number of edges is $n - 1$ for *n* vertices), commonly called *trees* [41]. Trees are graphs in which any two vertices are connected to each other by a finite path (connected graph) which can not contain cycles. Phrasing it like this makes it intuitively clear why this type of graph can represent a solution to e.g., decision trees or routing problems.

We distinguish vertices that are single end nodes (i.e., *leafs* in the tree) and those that are not (i.e., *branching* vertices). In the context of solutions, leaf nodes are e.g., outcomes of a process or represent empty sets of operations, while branching vertices represent choice points in a decision tree. There is only one node in the tree with no predecessor, called*root*. It is the initial starting node for the solution.

3.2   Complexities of Graphs

Given a set of *n* vertices, [10] showed that the family of different trees that can be constructed over this set has n^{n-2} members. This result is commonly known as *Cayley's Theorem* due to [13] (cf. [14]). The first combinatorial proof provided for this theorem was provided by Prüfer [43] in 1918 [41] using a mapping that represented trees with *n* vertices as strings of length $n - 2$ (cf. Sect. 3.3). By showing that this set of strings therefore had n^{n-2} members, Prüfer proved *Cayley's Theorem*.

If we restrict the branching factor for any vertex in the tree to a constant k, we get k-ary trees, which have been studied in the literature extensively [15, 23, 45]. The relation between *leaf* (n_{leaf}) and *branching* vertices n_{branch} in a k-ary tree is $n_{leaf} = n_{branch}(k-1) + 1$ (for an in depth discourse on k-ary tree graphs cf. [45]).

Given a maximum branching factor k and the total number n of nodes ($n = n_l + n_b$) we can calculate a minimum depth (vertices connecting the root with some leaf) h_{min} for the resulting trees: $h_{min} = log_k(n(k-1) + 1)$ [16]. Therefore, if $k = 2$ (the smallest possible branching factor) we get $h_{max} = log_2(n+1)$.

This number (h_{min}) gives us an insight into the minimum number of choice points required to reach the most distant leaf in tree. In other words, there is *at least* one path from the root to a specific leaf that will have a path of this length. Or, in cases where the path length is bounded, this dictates the minimum value for k.

Analogously, the maximum number of choices h_{max} in a tree is inversely proportional to the smallest value for k found in the solution: $h_{max} = log_2(n+1)$.

3.3 Encoding Graphs as Prüfer Code (PC)

In addition to providing a proof to [13], Prüfer also provided us with an efficient mechanism to encode trees into sequences of $n-2$ integers (and back); these so-called *Prüfer codes* (PC) are sequences of n elements taking values equal to the node indices. Effectively, Prüfer codes are $n-2$ dimensional Euclidean spaces, and these are known to work well with swarm and evolutionary search algorithms.

Since any tree can be uniquely represented by its Prüfer code [31], all possible spanning trees of n nodes are uniquely represented by all $(n-2)$-element Prüfer codes constructed from the set $\{1, \ldots, n\}$. Unfortunately, due to the specific way in which a Prüfer Code (PC) is constructed from a given tree graph, very similar PCs can represent fundamentally different trees [24] (see Fig. 1).

The fact that proximity in *Prüfer space* does not necessarily mean similarity when it comes to the corresponding tree graphs has been pointed out in the literature. This is generally considered as a fundamental problem of using PC to encode solutions for heuristics, e.g., locality is clearly not preserved. We address this issue in in Sect. 4, where we argue that, although Prüfer code has been shown to be a suboptimal choice as encoding for many optimization problems [24], they do work well in our case [4]. Our initial investigations are presented in [27].

3.3.1 Algorithms

Prüfer Code encoding and decoding follows a simple linear algorithm, details of which can be found in [31]. From e.g., [37] we know that there are $\Theta(n)$-complexity algorithms (i.e., algorithms that can perform the translation either way in linear time) to do this.

Algorithm 1 Encoding a tree-graph to Prüfer Code (cf. [37])

1: $L \leftarrow$ leafs of T
2: **for** $i \leftarrow 1$ to $(n-2)$ **do**:
3:     $v \leftarrow$ node removed from the head of L
4:     $PC[i] \leftarrow$ neighbour of v
5:     delete v from T
6:     **if** $deg(PC[i]) = 1$ **then**
7:         add $PC[i]$ to L

Note: the encoding algorithm (Algorithm 1) assumes that the leafs are stored in a list (initially sorted in ascending order).

Algorithm 2 Decoding Prüfer Code to a tree-graph (cf. [37])

1: $L \leftarrow$ nodes that do not appear in the Prüfer Code PC
2: **for** $i \leftarrow 1$ to $(n-2)$ **do**:
3:     $v \leftarrow$ node removed from the head of L
4:     add edge $\{v, PC[i]\}$ to T
5:     **if** i is the rightmost position of v in PC **then**
6:         add v to L
7:     $v \leftarrow$ node removed from the head of L
8:     add edge $\{v, PC[n-2]\}$ to T

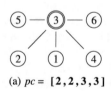

(a) $pc = [2,2,3,3]$

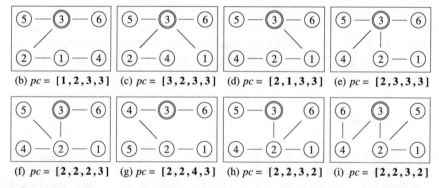

(b) $pc = [1,2,3,3]$ (c) $pc = [3,2,3,3]$ (d) $pc = [2,1,3,3]$ (e) $pc = [2,3,3,3]$

(f) $pc = [2,2,2,3]$ (g) $pc = [2,2,4,3]$ (h) $pc = [2,2,3,2]$ (i) $pc = [2,2,3,2]$

Fig. 1 The Prüfer codes similar to [2, 2, 3, 3] (**a**). All variations (**b**) to (**i**) differ from the original string (**a**) in only one digit and by only 1; the root vertex (double circle) is assumed to be v_3

3.3.2 Solution Space

Let's consider trees with n nodes (labelled 1 to n), resulting in PCs with $n - 2$ positions. We use $\mathbb{PC} = \{pc_1, \ldots, pc_{n^{(n-2)}}\}$ to denote the set of all possible PC that meet this description. Clearly, any $\mathbb{PC}$ can be mapped into a subset of $\mathbb{N}^+$ by reading individual pc_i as a number (e.g., for $n = 7$: this is $\{11111, \ldots, 26416, 26417, 26421, \ldots, 77777\}$). We use a PC's position in this set as the its ID (see example).

When exploring the solution space with heuristics we want there to be some correlation between the location of a solution in that space and its performance value. If we require that *similar* PCs represent trees that encode similar solutions, we have to consider how we define *similar*.

Example: Let's consider encoding cooking recipes as trees (representing the order and inter-dependency of individual steps, started with step v_1). For a recipe with 7 steps, this can be represented as a tree with 7 nodes (of which there are exactly 16807 unique variations), each corresponding to exactly one PC with 5 positions.

If, in this example, the definition of *similar* is simply the numerical distance between two codes (e.g., 24617 is followed immediately by 26421, cf. Fig. 2 bottom row) then very similar PCs encode substantially different trees. As pointed out in [24] this will make PC a sub-optimal choice for interpretations of similarity.

4 Navigating Prüfer Code

While the variations shown in Fig. 2 differ, they do not differ *dramatically*. As it turns out, this loose similarity was already enough to empower a PSO approach to produce results of sufficient quality (optimal or near-optimal, actually) when we used PC to encode solutions representing cable diagrams [3, 4, 26].

The way trees are constructed from PC (cf. Algorithm 2) implies that the connectivity of a vertex (the number of vertices it is connected to) is equal to the number of its occurrences in the PC + 1. This also means that every non-occurring vertex is a leaf. On the other hand, the position of the integers matters, and exchanging two integers can result in more than the exchange of the corresponding vertices in the tree (cf. Fig. 2). In the next section we will discuss how these simple insights allow us to impose filters on $\mathbb{PC}$, resulting in a number of sub-solution spaces with intuitively obvious ordering that capture properties which we were specifically interested in.

Using the above insight we look at certain filters for PCs that characterize properties of interest to us. These filters, are defined with respect to a specific $pc_i \in \mathbb{PC}$:

- $\mathbb{I}_{pc_i}$, the set of all different integers that occur in pc_i
- $\mathbb{I}_{pc_i}^+$, the ordered list of all the occurring integers

This defines sets from which we can create subsets of $\mathbb{PC}$. The idea is that we can impose a property preserving ordering on subsets of the solution space.

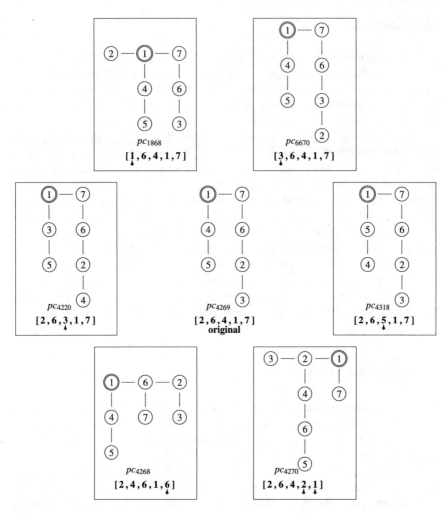

Fig. 2 Variations on pc_{4269}. In each row, the middle graph is the original and the outside graphs are variations that are neighbours: (first row) differing in one position and just by 1 from the original integer (solution space $(n - 2)$-dimensional) or (second row) the previous and the next ID (solution space: $\mathbb{N}^+$); the root vertex is assumed to be v_1

Example: for trees with $n = 5$, $\mathbb{PC} = \{[1, 1, 1], \ldots, [5, 5, 5]\}$; for e.g., $pc_i =$ [1, 2, 1]:

- $\mathbb{I}_{[1,2,1]} = \{1, 2\}$ and
- $\mathbb{I}^+_{[1,2,1]} = \{1, 1, 2\}$.

We use these filters to define the similarity classes $\mathbb{PC}_{\mathbb{I}}$ and $\mathbb{PC}_{\mathbb{I}^+}$, i.e., the subsets of $\mathbb{PC}$ where all members pc_j have the same $\mathbb{I}_{pc_j}$ or $\mathbb{I}^+_{pc_j}$, respectively:

- $\mathbb{PC}_{\mathbb{I}_{pc_i}}$, the subset of $\mathbb{PC}$ in which members are constructed using only the integers found in pc_i, and
- $\mathbb{PC}_{\mathbb{I}^+_{pc_i}}$, where all members have exactly the same integers as pc_i, but not necessarily in the same order.

This means that : $\forall pc_j \in \mathbb{PC}_{\mathbb{I}_{pc_i}} : \mathbb{I}_{pc_i} = \mathbb{I}_{pc_j}$ and $\forall pc_k \in \mathbb{PC}_{\mathbb{I}^+_{pc_i}} : \mathbb{I}^+_{pc_i} = \mathbb{I}^+_{pc_k}$. Note that pc_i is always included: $pc_i \in \mathbb{PC}_{\mathbb{I}_{pc_i}}$ and $pc_i \in \mathbb{PC}_{\mathbb{I}^+_{pc_i}}$; furthermore $\mathbb{PC}_{\mathbb{I}^+_{pc_i}} \subset \mathbb{PC}_{\mathbb{I}_{pc_i}}$.

Example: for $\mathbb{I}_{[1,2,1]} = \{1, 2\}$ and $\mathbb{I}^+_{[1,2,1]} = \{1, 1, 2\}$ we get:

- $\mathbb{PC}_{\mathbb{I}_{[1,2,1]}} = \{[1, 1, 2], [1, 2, 1], [1, 2, 2], [2, 1, 1], [2, 1, 2], [2, 2, 1], [2, 2, 2]\}$ and
- $\mathbb{PC}_{\mathbb{I}^+_{[1,2,1]}} = \{[1, 1, 2], [1, 2, 1], [2, 1, 1]\}$.

To create – individually for each pc_i – relative pc_i-solution spaces based on $\mathbb{I}_{pc_i}$ or $\mathbb{I}^+_{pc_i}$ we need to define a distance between pc_i and any pc_j in $\mathbb{PC}_{\mathbb{I}_{pc_i}}$ and $\mathbb{PC}_{\mathbb{I}^+_{pc_i}}$. Clearly the distance to itself ($pc_j = pc_i$) is zero, and if we define the set of all pc_j with a distance equal to one (i.e., the set of immediate neighbours of pc_i) then we can calculate the distance $\delta(i, j)$ between any two pc_i and pc_j as the shortest path connecting these two through their neighbours.

We may either want a to define a single neighbour, a certain number of neighbours or sets of neighbours (potentially of varying sizes). This will directly impact the dimensionality of our solutions space: with a single neighbour we can use $\mathbb{N}^+$ as solution space, otherwise our solution space is n-dimensional.

Example: for both $\mathbb{PC}_{\mathbb{I}_{pc_i}}$ and $\mathbb{PC}_{\mathbb{I}^+_{pc_i}}$ neighbourhood could (the choice is problem specific) be defined as, e.g.:

- the element in the respective set that is numerically the closest to pc_i (reading e.g., $[1, 3, 2, 4]$ as 1324), or
- all those elements that are created by exchanging two neighbouring digits of the pc, e.g., for $pc_i = [1, 2, 3, 4]$ this would be $[2, 1, 3, 4]$, $[1, 3, 2, 4]$ and $[1, 2, 4, 3]$.

When optimizing cabling structures for e.g., distributed antenna systems or routing network trees, the number of used *splitters* or *routers* (corresponding to branches in the tree) is an important factor as hardware plays a major role in the overall cost. In problems of this type constraints are commonly imposed on all paths from the root to the leaf nodes of the trees (e.g., power attenuation due to cable length which must not exceed a certain value).

On the other hand, having identified nodes in the network that exhibit high potential to become branches we want to consider changing their branching factors (i.e., the equivalent of replacing a splitter with a larger or a smaller one).

Specifically, our subsets of $\mathbb{PC}$ allow us the following:

1. $\forall pc_i \in \mathbb{PC}_{\mathbb{I}^+_{pc_{original}}}$: pc_i preserves the number of branching nodes, their branching degree as well as which node has how many branches. Only the specific allocation of leafs to these branches changes, as well as how these branching nodes are connected to each other.

2. $\forall pc_j \in \mathbb{PC}_{\mathbb{I}_{pc_{original}}}$: contrary to pc_i above, pc_j does not ensure that the number of nodes with a certain branching degree stays the same, i.e., while the branching nodes do not change, their degree might, as does (as above) which leafs/other branching nodes they connect to.
3. In addition to the two above, we can explore variations on $\mathbb{PC}_{\mathbb{I}_{pc_{original}}}$ and $\mathbb{PC}_{\mathbb{I}^+_{pc_{original}}}$ by replacing all occurrences of an integer with one that does not occur in the original, or by simply adding or removing integers. As shown in Fig. 1, these are more dramatic changes.

5 Proof of Concept Through Application

Despite the claims made in [24] we have adopted Prüfer Code as the encoding for solutions to the problem of planning the cabling to power indoor antenna systems for large buildings [3] (cf. Chap. 6 for a modelling). Table 1 indicates the complexity of the problem, for the example cost of 1 s per 1000 evaluated solutions.

Figure 3 illustrates the problem and shows that its solutions can be represented by trees. Our work, which was tested for hypothetical problems of up to 100 floors

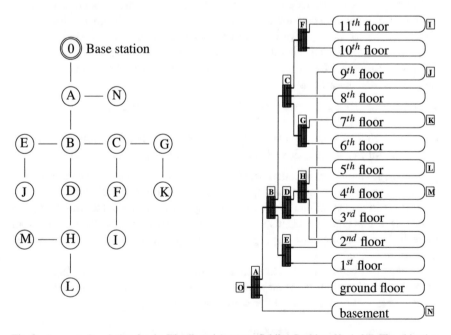

Fig. 3 An example solution for the Distributed Antenna Cabling Problem [3, 4, 26]. The objective is to connect all floors (and antennas on each floor) using splitters and cables, subject to power constraints imposed in the splitters and the antennas. The choice of branching nodes (and their degree) is a primary factor in this problem, making Prüfer Code a useful encoding

Table 1 The number of distinct trees with n nodes. The brute force approach of evaluating all (n^{n-2}) different solutions becomes computationally unfeasible for moderate values for n

n	n^{n-2}	if it takes 1 s to evaluate 1000 trees
5	125	< 1 s
7	16,807	≈17 s
8	262,144	≈4.4 min
10	1.0×10^8	≈28 days
15	1.94×10^{15}	≈ 61,713 years
20	2.62×10^{22}	≈8.3 $\times 10^{12}$ years

Fig. 4 The probability of improving on the current best known solution, plotted against the number of PSO generations without improvement. If the encoding of solutions into prüfer space were void of any underlying structure relating *good/better* solutions to one another, then the PSO could not – reliably – find a better solution within 10 generations (or never thereafter). This directly implies that PC does preserve relevant properties sufficiently

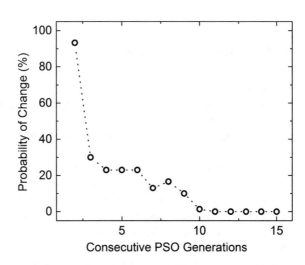

(cf. Chap. 6 for details), showed that using Particle Swarm Optimization to explore a PC solution space resulted in good solutions in short time.

The approach was compared to an implementation using Genetic Algorithms to solve the problem (using the same encoding into Prüfer Code solution space) and, while GA were inferior to our PSO approach, both performed well, indicating that using PC was a feasible approach. The interested reader is referred to [4] for an overview and to Chap. 6 for details on the obtained results.

Furthermore, our investigations into the performance of the algorithm showed that the PSO does converge towards good solutions. Amongst others, this is indicated by the fact that stagnation in the improvement over previous generations indicates having found the best expectable solution (cf. Fig. 4).

The argument is straight forward: if our exploration through PC-space were entirely random (in other words: void of beneficial *similarities*) we would expect that the potential for finding improved solutions increased with additional searches, while the graph plotted in Fig. 4 indicates the opposite.

The solutions returned by our PSO implementation showed a large number of individual solutions with very similar good performance values. In other words, our PSO identified clusters of solutions sharing performance relevant properties by converging on the respective representations in Prüfer Code. The similarity classes we proposed in Sect. 4 may be part of what is going on here, or the good performance of our system may indicate that the PSO has learned the structure underlying the distribution of *similar* trees (similar in the context of our I-DAS problem) when mapped to the corresponding PC solution space. This, however, is mere conjecture and the subject of further investigations. What we can state with certainty is that choosing PC to encode the solution space for a heuristic is not always a bad idea. If it were, we could not have obtained the very good results in our case.

6 Conclusion and Context

It is important to emphasises that our investigations and the suggestions put forward in this chapter are not meant to be a rebuttal to the claims made in [24]. Instead, they are to be understood as an addition, in the sense that the we provide a complex problem (together with a heuristic-based approach to solve it) where the encoding of trees in PC is resulting in very good performance of the system.

On the basis of this we argue that using Prüfer Code encoding *not always* a bad idea. Specifically, when using trees to represent (a) variations on the branching of a tree (both in identifying the branching nodes we well as their degree of branching) and (b) the allocation of leaf nodes to branching nodes, Prüfer Code has proven to be a useful encoding. We intend to investigate this further by using it as encoding for other problems/problem classes as well as through a more formal analysis.

There have been other investigations into locality properties of PC (e.g., [41]) which also hint that the general results of [24] may not be all there is to PC. We have a number of conjectures about this, which will require extensive further investigations and are currently outside the scope of this chapter.

Acknowledgements The authors are grateful for the support from the UAE ICT-Fund on the project *"Biologically Inspired Network Services"*. We acknowledge K. Poon (EBTIC, KUST) for bringing the I-DAS problem to our attention. HH acknowledges the hospitality of the EBTIC Institute and F. Saffre (EBTIC, KUST) during his fellowship 2017.

References

1. Almeida, M., Hildmann, H., Solmazc, G.: Distributed UAV-swarm-based real-time geomatic data collection under dynamically changing resolution requirements. In: UAV-g 2017 - International Conference on Unmanned Aerial Vehicles in Geomatics, ISPRS Archives of the Photogrammetry, Remote Sensing and Spatial Information Sciences, Bonn, Germany (September 2017)

2. Anderson, C., Boomsma, J.J., Bartholdi, J.J.: Task partitioning in insect societies bucket brigades. Insectes Soc. **49**, 171–180 (2002)
3. Atia, D.Y.: Indoor distributed antenna systems deployment optimization with particle swarm optimization. M.Sc. thesis, Khalifa University of Science, Technology (2015)
4. Atia, D.Y., Ruta, D., Poon, K., Ouali, A., Isakovic, A.F.: Cost effective, scalable design of indoor distributed antenna systems based on particle swarm optimization and prufer strings. In: IEEE 2016 Congress on Evolutionary Computation, Vancouver, Canada. IEEE, New York (July 2016)
5. Bartholdi, J.J., Eisenstein, D.D.: A production line that balances itself. Oper. Res. **44**(1), 21–34 (1996)
6. Berdahl, A., Torney, C.J., Ioannou, C.C., Faria, J.J., Couzin, I.D.: Emergent sensing of complex environments by mobile animal groups. Science **339**(6119), 574–576 (2013)
7. Blackburn, P., deRijke, M., Venema, Y.: Modal Logic. Cambridge University Press, Cambridge (2001)
8. Bonabeau, E., Dorigo, M., Theraulaz, G.: Swarm Intelligence: From Natural to Artificial Systems. SFI Studies on the Sciences of Complexity. Oxford University Press, New York (1999)
9. Bonabeau, E., Dorigo, M., Theraulaz, G.: Inspiration for optimization from social insect behaviour. Nature **406**, 39–42 (2000)
10. Borchardt, C.W.: über eine Interpolationsformel für eine Art symmetrischer Funktionen und über deren Anwendung. In: Math. Abh. Akad. Wiss. zu Berlin, pp. 1–20. Berlin (1860)
11. Brownlee, J.: Clever Algorithms: Nature-inspired Programming Recipes. www.Lulu.com (2011)
12. Camazine, S., Deneubourg, J.-L., Franks, N.R., Sneyd, J., Theraulaz, G., Bonabeau, E.: Self-organization in Biological Systems. Princeton University Press, Princeton (2001)
13. Cayley, A.: On the theory of the analytical forms called trees. Phil. Mag. **13**, 172–176 (1857)
14. Cayley, A.: volume 13 of Cambridge Library Collection - Mathematics, p. 2628. Cambridge University Press, Cambridge (July 2009)
15. Cha, S.-H.: On complete and size balanced k-ary tree integer sequences. Int. J. Appl. Math. Inf. **6**(2), 67–75 (2012)
16. Cha, S.-H.: On integer sequences derived from balanced k-ary trees. In: Proceedings of American Conference on Applied Mathematics, pp. 377–381. Cambridge, MA (January 2012)
17. Diestel, R.: Graph Theory. Electronic Library of Mathematics. Springer, Berlin (2006)
18. Dorigo, M., Stützle, T.: Ant Colony Optimization. Bradford Company, Scituate (2004)
19. Ducatelle, F., Di Caro, G.A., Gambardella, L.M.: Principles and applications of swarm intelligence for adaptive routing in telecommunications networks. Swarm Intell. **4**(3), 173–198 (2010)
20. Fraser, A., Burnell, D.G.: Computer Models in Genetics. McGraw-Hill, New York (1970)
21. Fraser, A.S.: Simulation of genetic systems by automatic digital computers 1. Introduction. Aust. J. Biol. Sci. **10**, 484–491 (1957)
22. Gabrys, B., Ruta, D.: Genetic algorithms in classifier fusion. Appl. Soft Comput. **6**(4), 337–347 (2006)
23. Ghosh, S.K., Ghosh, J., Pal, R.K.: A new algorithm to represent a given k-ary tree into its equivalent binary tree structure. J. Phys. Sci. **12**, 253–264 (2008)
24. Gottlieb, J., Julstrom, B.A., Raidl, G.R., Rothlauf, F.: Prüfer numbers: a poor representation of spanning trees for evolutionary search. In: Proceedings of the Genetic and Evolutionary Computation Conference (GECCO 2001), San Francisco, CA, USA, pp. 343–350. Morgan Kaufmann Publishers, Burlington (2001)
25. Halloy, J., Sempo, G., Caprari, G., Rivault, C., Asadpour, M., Tâche, F., Saïd, I., Durier, V., Canonge, S., Amé, J.M., Detrain, C., Correll, N., Martinoli, A., Mondada, F., Siegwart, R., Deneubourg, J.L.: Social integration of robots into groups of cockroaches to control self-organized choices. Science **318**(5853), 1155–1158 (2007)
26. Hildmann, H., Atia, D.Y., Ruta, D., Poon, K., Isakovic, A.F.: Nature-inspired optimization in the Era of IoT: Particle Swarm Optimization (PSO) applied to Indoor Distributed Antenna Systems (I-DAS), chapter TBD, page TBD. Springer, Berlin (2018) (forthcoming)

27. Hildmann, H. Ruta, D., Atia, D.Y., Isakovic, A.F.: Using branching-property preserving Prüfer code to encode solutions for particle swarm optimisation. In: 2017 Federated Conference on Computer Science and Information Systems (FedCSIS), pp. 429–432 (September 2017)
28. Hildmann, H., Martin, M.: Resource allocation and scheduling based on emergent behaviours in multi-agent scenarios. In: Vitoriano, B.na, Parlier, G.H. (Eds.) Proceedings of the International Conference on Operations Research and Enterprise Systems, pp. 140–147, Lisbon, Portugal (January 2015). INSTICC, SCITEPRESS
29. Hildmann, H., Sebastien Nicolas, A.: self-organizing client, server allocation algorithm for applications with non-linear cost functions. In: IEEE PES Innovative Smart Grid Technologies Latin America (2015 ISGT-LA), p. 2015. Montevideo (October 2015)
30. Hildmann, H., Nicolas, S., Saffre, F.: A bio-inspired resource-saving approach to dynamic client-server association. IEEE Intell. Syst. **27**(6), 17–25 (2012)
31. Julstrom, B.A.: Exercises in data structures, quick decoding and encoding of prüfer strings (2005)
32. Kennedy, J., Eberhart, R.: Particle swarm optimization. In: Proceedings of IEEE International Conference on Neural Networks, vol. 4, pp. 1942–1948 (November 1995)
33. Kuila, P., Jana, P.K.: Energy efficient clustering and routing algorithms for wireless sensor networks: particle swarm optimization approach. Eng. Appl. Artif. Intell. **33**, 127–140 (2014)
34. Lim, S., Rus, D.: Stochastic distributed multi-agent planning and applications to traffic. In: ICRA, pp. 2873–2879. IEEE, New York (2012)
35. Ma, R.-J., Yu, N.-Y., Hu, J.-Y.: Application of particle swarm optimization algorithm in the heating system planning problem. Sci. World J. (2013)
36. Macaš, M., Gabrys, B., Ruta, D., Thotská, L.: Particle swarm optimization of multiple classifier systems. In: 9th International Work-Conference on Artificial Neural Networks, pp. 333–340 (2007)
37. Micikevičius, P., Caminiti, S., Deo, N.: Linear-time algorithms for encoding trees as sequences of node labels (2007)
38. Mitri, S., Floreano, D., Keller, L.: The evolution of information suppression in communicating robots with conflicting interests. Proc. Nat. Acad. Sci. **106**(37), 15786–15790 (2009)
39. Mugler, A., Bailey, A.G., Takahashi, K., ten Wolde, P.R.: Membrane clustering and the role of rebinding in biochemical signaling. Biophys. J. **102**(5), 1069–1078 (2012)
40. Navlakha, S., Bar-Joseph, Z.: Algorithms in nature: the convergence of systems biology and computational thinking. Mol. Syst. Biol. **7**, 546 (2011)
41. Paulden, T., Smith, D.K.: Developing new locality results for the prüfer code using a remarkable linear-time decoding algorithm. Elect. J. Comb. **14**(1) (August 2007)
42. Pearl, J.: Heuristics: Intelligent Search Strategies for Computer Problem Solving. Addison-Wesley, The Addison-Wesley Series in Artificial Intelligence (1984)
43. Prüfer, H.: Neuer Beweis eines Satzes über Permutationen. Archiv der Mathematik und Physik **27**, 742–744 (1918)
44. Raman, S., Raina, G., Hildmann, H., Saffre, F.: Ant-colony based heuristics to minimize power and delay in the internet. In: IEEE International Conference on Green Computing and Communications 2013 (IEEE GreenCom 2013 WS - Greencom-Next 2013), Beijing, P.R. China
45. Ramanan, P.V., Liu, C.L.: Permutation representation of k-ary trees. Theor. Comput. Sci. **38**, 83–98 (1985)
46. Renfrew, D., Yu, X.H.: Traffic signal control with swarm intelligence. In: 2009 Fifth International Conference on Natural Computation, vol. 3, pp. 79–83 (August 2009)
47. Saffre, F., Furey, R., Krafft, B., Deneubourg, J.-L.: Collective decision-making in social spiders: dragline-mediated amplification process acts as a recruitment mechanism. J. Theor. Biol. **198**, 507–517 (1999)
48. Saffre, F., Hildmann, H., Deneubourg, J.-L.: Can individual heterogeneity influence self-organised patterns in the termite nest construction model? Swarm Intell. (October 2017)
49. Schoonderwoerd, R., Bruten, J.L., Holland, O.E., Rothkrantz, L.J.M.: Ant-based load balancing in telecommunications networks. Adapt. Behav. **5**(2), 169–207 (1996)

50. Taylor, J.G., Cutsuridis, V., Hartley, M., Althoefer, K., Nanayakkara, T.: Observational learning: basis, experimental results and models, and implications for robotics. Cogn. Comput. **5**(3), 340–354 (2013)
51. Trelea, I.C., Ioan Cristian Trelea: The particle swarm optimization algorithm: convergence analysis and parameter selection. Inf. Process. Lett. **85**(6), 317–325 (2003)
52. Turing, A.M.: Computing machinery and intelligence (1950)
53. Werfel, J., Petersen, K., Nagpal, R.: Designing collective behavior in a termite-inspired robot construction team. Science **343**(6172), 754–758 (2014)
54. Zhou, L., Li, B., Wang, F.: Particle swarm optimization model of distributed network planning. JNW **8**(10), 2263–2268 (2013)

A Model for Wireless-Access Network Topology and a PSO-Based Approach for Its Optimization

H. Hildmann, D. Y. Atia, D. Ruta, S. S. Khrais and A. F. Isakovic

Abstract By the year 2020, the global network of connected sensors and devices will contain 50 billion connected devices and be the single largest factor in global power consumption. The planet's ICT infrastructure already exceeds 10% of mankind's power consumption (tendency: rising). The complexity of designing the topology for extend wireless access to ensure a thorough and economically sound signal coverage in buildings (from a building's base station to distributed antennas throughout the building, through a complex network of coaxial cables and power splitters) increases exponentially ($O(n^{n-2})$). We present our results from using Particle Swarm Optimization (PSO) to provide near optimal network topology for distributed in-building antenna systems. We use Prüfer code representation to efficiently traverse through different spanning tree solutions. Our approach is scalable and robust, capable of producing I-DAS design advice for buildings beyond one hundred floors. We demonstrate that our model is capable of obtaining optimal solutions for small buildings and near optimal solutions for tall buildings.

H. Hildmann (✉)
Dep. de Ingeniería de Sistemas y Automática, Universidad Carlos III de Madrid (UC3M),
Av. Universidad, 30, 28911 Léganes, Spain
e-mail: hanno.hildmann@uc3m.es

D. Y. Atia · S. S. Khrais
Khalifa University of Science and Technology, 127788 Abu Dhabi, UAE
e-mail: dina.atia@ku.ac.ae

S. S. Khrais
e-mail: sena.khrais@gmail.com

D. Ruta
EBTIC, Khalifa University of Science and Technology, 127788 Abu Dhabi, UAE
e-mail: dymitr.ruta@ku.ac.ae

A. F. Isakovic
Physics Department, Khalifa University of Science and Technology,
127788 Abu Dhabi, UAE
e-mail: iregx137@gmail.com; abdel.isakovic@ku.ac.ae

© Springer Nature Switzerland AG 2019
S. Fidanova (ed.), *Recent Advances in Computational Optimization*,
Studies in Computational Intelligence 795,
https://doi.org/10.1007/978-3-319-99648-6_6

87

1 Introduction

In recent years the so-called Internet of Things (IoT) has increasingly received the attention of the media and the academic world, as evidenced by the growing number of occurrences of the term in main print and digital media or the surge in funding to IoT related projects, respectively. Looking at the technical side of this, the IoT is a rapidly growing network of sensors and devices which are increasingly integrated into virtually all types of objects of our daily lives, many of which are relying on wireless access networks. To put this in numbers: by the year 2020, this network will contain 50 billion connected devices and the global IoT/Machine to Machine (M2M) communications market will have a volume of $0.5 trillion [22].

Wireless access networks – though only a part of the worlds ICT systems – are steadily increasing in size as well as importance and constitute a growing part of the systems power consumption. The global ICT is already accounting for approximately 10% of the global power consumption, putting it at about 150% of the demand of the global aviation industry or, in other terms, rivaling the combined consumption of e.g. Germany and Japan.

We propose an approach for energy-efficient and power-consumption aware design of indoor wireless access networks: In-building Distributed Antenna Systems (I-DAS). These systems are commonly rolled out across entire buildings and can comprise of dozens or hundreds of antennas (cf. Fig. 1); however, the optimal configuration of the underlying cabling is unfortunately of a complexity that makes the deterministic calculation of the optimal solution computationally unfeasible for all but the smallest networks. Our approach is scalable, robust and capable of generating I-DAS topologies for buildings beyond one hundred floors. Our results show that our model produced near optimal solutions in extremely short time.

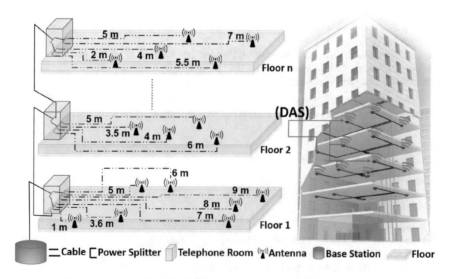

Fig. 1 An illustration of an In-building DAS and the components typically used

2 (Indoor) Distributed Antenna Systems (I-DAS)

Distributed antenna systems (DAS) are a promising approach to address the challenge of supplying a rapidly increasing demand for bandwidth to a growing number of mobile devices. DAS offer high spectral efficiency and provide good wireless access coverage over an area [21]. Contrary to the more traditional approach which relies on a single antenna to cover a large area, DAS utilizes a number of antennas, distributed throughout the area that needs coverage. Each antenna is connected by cable to a central device [19]. This offers high spectral efficiency for wireless communication [15] and reduces correlations due to spatially separated antennas [20].

The potential benefits that DAS and the indoor DAS (henceforth I-DAS) may offer over the traditional one-antenna approach have been studied analytically in e.g. [12] and results indicate that dedicated I-DAS can consistently deliver high data throughput and good signal quality to large numbers of indoor users. There is, however, the issue of network throughput degradation due to performance bottlenecks in periods of high usage from many users in the building. Beijner [5] discusses this in the context of using dedicated I-DAS to mitigate such factors when providing indoor wireless access network coverage and concluded that in-building DAS are a dominant solution for offering in-building coverage, especially for 3G networks.

So if we accept the above and agree that I-DAS may offer advantages, what then are the challenges in using them? We'd like to distinguish two areas of interest:

1. the design and deployment (one time investment, CapEx) of such a system, and
2. the operation thereof (continuously incurred costs, OpEx).

The former includes the cost of the hardware, the positioning of the antennas to minimize redundancy and the challenge of designing the cabling layout optimally, the latter is affected by the e.g. resulting power requirements and power leakage of the I-DAS. These challenges are reflected by the investigations (and proposed solutions) found in the literature where a number of different, and potentially separate, opportunities for improvement have been investigated. We present a few examples here but for a more detailed literature study the reader is referred to [3].

The authors of [2] use genetic algorithms (GA) to optimize the locations and total number of antennas as well as their individual transmitting power. They showed (albeit on a relatively small DAS network) that number and location of antennas, as well as the power transmitted from the respective antennas, can be optimized by varying the transmitted power for individual antennas. In addition, they achieved a reduced power leakage of the entire DAS. Their results indicate that non-homogeneous transmission levels have the potential to reduce the number of antennas used as well as the power leakage incurred by them. Optimization of the underlying cabling structure has been addressed by e.g. Mixed Integer Linear Programming (MILP) [9], providing optimal cabling solutions on a floor by floor level (as opposed to for the entire building across all floors). Their model used topology constraints imposed on the building and focused on the optimal usage of splitters under consideration of the antenna power requirements.

I.e., the approach can be described as prioritizing the reduction of cable length and thus the reduction of the deployment cost. Note that we argue that solving the problem for floors independently means that the approach does not scale well; the approach suggested in this article offers a solution for entire buildings with large numbers of floors. This concern notwithstanding, the available literature suggests that there are substantial gains to be made already by optimizing the hardware deployment alone. In addition to optimizing antenna location and transmission output and assuming an already optimized DAS tree topology (a tree structure with the base station at the root and antennas at the leaves) in place, splitters have to be used as branches in the topology tree. Calculating the allocation of specific cables to specific splitters (and determining the optimal splitter type for such an allocation) is important as the surplus between the required and the received power (i.e. the antenna power deviation) at each splitter impacts the operational cost of the I-DAS. In the literature, e.g. [1] proposed exact pseudo-polynomial time algorithms and fully-polynomial time approximation schemes to determine optimal/near optimal allocation of splitters to base station (inner node of the cabling tree) as well as antenna to splitter (leaf to inner node). This enabled them to minimize the antenna power deviation (i.e. the difference between the required and the received power). When compared to standard approaches their implementation outperformed greedy algorithms. The DAS/I-DAS problem can be modelled as a non-cyclic graph, commonly called a *tree* [17]. There are n^{n-2} variations on trees with n nodes [6]. A DAS topology can be represented as a connected graph without cycles (i.e. a tree, [17]): all n antennas in an area are represented by the n nodes in a graph, in which any two nodes are connected to each other (connected graph) and where the number of edges is $n - 1$ with n the number of nodes (i.e. there are no cycles). This result is commonly known as *Cayley's Theorem* due to [7] (cf. [8]).

Assuming that it takes 1 s to evaluate 1000 different solutions (trees representing unique variations on a cabling topology for a DAS or I-DAS problem), Table 1 illustrates the complexity of the I-DAS problem by estimating computation times to find the optimal solution.

Table 1 The computational complexity of determining the optimal configuration for a DAS problem (assuming that 1000 solution candidates can be evaluated computationally in 1 s)

Antennas	Unique solutions	Time to compute the optimal solution
5	125	<1 s
7	16,807	≈17 s
8	262,144	≈4.4 min
10	1.0×10^8	≈28 days
15	1.94×10^{15}	≈61,713 years
20	2.62×10^{22}	$\approx 8.3 \times 10^{12}$ years

3 Models

The purpose of an In-Building Distributed Antennas System (I-DAS) is to provide cellular coverage of varying throughput to different areas of a building. Solutions to this can differ greatly in the resulting one-time equipment and installation cost as well as the continuing costs incurred for operating and maintaining the I-DAS.

An I-DAS consists of a center base station supplying power to all the floors using coaxial cables routed through each floor's telephone room. It also contains passive splitters to split the incoming signal into multiple output ports. Figure 1 in the introduction shows the components of an I-DAS problem. The underlying structure of such an (indoor) distributed antenna system (a single point of origin from which cables branch out to either antennas or to passive splitters, which in turn have multiple cables branching out from them) suggest to model the system as a connected graph without cycles. Such graphs are commonly referred to as *trees* [17].

3.1 Assumptions and Simplifications

The I-DAS problem for a building will concern F, a set of individual floors f_i. We assume a set A of all antennas a and for our model we assume that all antennas in A are actually used in our solution. We furthermore assume a set S of splitters (and again, we only include the splitters that are eventually used in the deployment): there are a number of different types of splitters and we write S_t (with $S_t \subseteq S$) for the set of splitters of type t (i.e. the set of 't-way splitters'), with the different types are mutually exclusive (i.e. partitioning S: for any two $S_{t_i}, S_{t_j} \in S : S_{t_i} \cap S_{t_j} = \emptyset$).

As far as the hardware is concerned we assume that only 4 types of splitters are used (differing in the number of their output ports): 2-way, 3-way, 4-way and 6-way splitters. The incurred power loss of each of these is assumed to be of 3.0, 4.8, 6 and 7.8 dBm, respectively. Given our experience with deployments of I-DAS in the real world we model a system such that all splitters on a floor are located in one location (the *telephone room*). In other words, the splitters connecting the antennas on a specific floor as well as the splitters that connect all floors to the base station (cf. Fig. 1 in the Introduction) are physically co-located in the telephone rooms on their respective floors. The antenna distribution on a floor is assumed to be pre-determined (and optimized) which means that the cost for deployment of the cabling and cost thereof is also fixed.

However, the decision of which antennas to connect to which splitter is subject to optimization, which operates under the constraints of antenna-specific minimum power requirements that need to be satisfied to ensure each antenna can operate as required for their predetermined location. Note that we assume that tapers or attenuators can be used to limit the leakage in case of overpowering. The choice of which splitter type to use and which antennas to connect to a specific splitter has the potential to significantly impact the respective splitters' power requirements.

I-DAS are normally designed modular: in a building with a set F of floors, the I-DAS topology on each floor f_i ($f_i \in F$) is self-contained with a single connection to the rest of the building. Because of this, the telephone rooms on all floors constitute a single entry node into floor f_i, the power requirement of which is equal to the requirement of the one splitter from which the entire floor is powered. Owing to this we have a second level of optimization (the connection of all floors to the base station in the building). This means that our model of the problem effectively enables us to treat the I-DAS problem as two independent I-DAS optimization problems:

1. **Intra-Floor optimization**, we assumes fixed antennas locations and optimizes only the topology (i.e. connecting the splitters and antennas) on a single floor.
2. **Inter-Floor optimization**, optimizes the building's I-DAS topology (only the challenge of connecting the base station to the telephone rooms on all floors).

Using this distinction we only consider the antennas in set A when we are optimizing the individual floors, where we require for each antenna a information about the power requirements of the antenna ($power_a$) and how much cable is required to connect this antenna to the telephone room ($dist_a$). Regarding the hardware cost, we ignore the cost of the antennas since they are fixed and not subject to our optimization. We do, however, consider the cost for splitters ($\text{cost}_{splitter_t}$, which may differ for different splitter types) as well as for the cabling (cost_{cable}). Finally, we have to factor the power attenuation incurred by splitters ($\text{att}_{splitter_t}$, again, different for different splitter types) as well as by the cabling (att_{cable}) into our calculations (Fig. 2).

Since we are treating the Inter-Floor optimization problem and the Intra-floor optimization problem separately we effectively impose a second partitioning on S:

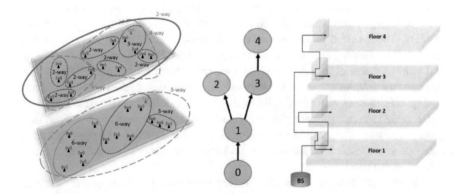

Fig. 2 A visual representation of the 2- and 6-way splitter tree algorithm (cf. Sect. 4.2 on page 98) for the Intra-Floor Problem (left) and an example of a solution to an Inter-Floor Problem with the spanning tree (middle) and the actual splitter structure (right) connecting floors in a 4-levels building. By solving the topology of each floor separately and disconnecting the inter-floor and the intra-floor problems, we greatly reduce the complexity of a problem

- for each floor f_i there is a set S_{f_i} of splitters located on this floor but only a single splitter s_{f_i} that connects the floor to the rest of the building, called *entry-node*.
- for the entire building, there is a set S_F of splitters connecting the main output splitter (called s_F) to the entry-nodes s_{f_i} in the telephone rooms on all floors.

Therefore, S can be partitioned as $S_{f_i} \cup S_F$, and this partition is mutually exclusive (i.e. $S_{f_i} \cap S_F = \emptyset$). Determining this partitioning is part of finding a solution.

We argue that this conceptual separation allows us to scale our approach to solve I-DAS problems for buildings with a large numbers of floors. In what follows we present the mathematical models used for both optimization problems.

3.2 Solving the Problem at the Floor Level (Intra-Floor)

3.2.1 Intra-Floor Hardware Cost

Since approaches to determine optimal deployment for a specific floor layout and under a given set of service requirements are readily available in the literature, we assume this (the specification of the number of antennas as well as the optimal locations for them, on any given floor) as given. This allows us to compartmentalize the problem. We can use the number of antennas as an input to our approach and entirely ignore the cabling required to connect them to the telephone room, since this is going to be constant across any possible topology for a specific floor). Furthermore, since the splitters are physically co-located in the telephone room we take the liberty and ignore the small amount of cable required to interconnect the splitters, arguing that this will be a negligible expense.

The hardware cost $Cost_{CapEx}^{f_i}$ for a single floor f_i is thus given by:

$$Cost_{CapEx}^{f_i} = \sum_{\substack{t \in \{2,3,4,6\} \\ s \in S_t}} \text{cost}_{splitter_t} \times \tau_s^i + \sum_{a \in A} dist_a \times \text{cost}_{cable} \times \tau_a^i \qquad (1)$$

With S_t the set of all splitters of type t (i.e. the set of '$t - waysplitters$'), τ_s^i a boolean that has the value 1 *iff* splitter s is located on floor f_i, A the set of all antennas, $dist_a$ the distance of the antenna a to the telephone room and τ_a^i a boolean that has the value 1 *iff* antenna a is located at floor f_i. The constants cost_{cable} and $\text{cost}_{splitter_t}$ are the cost per meter cable and for a splitter of type t, respectively.

3.2.2 Intra-Floor Operational Cost

Similarly, the operational cost for floor f_i has to take *Att.cable.ant*$_a^i$ (the power attenuation of the cabling, using a constant per meter, att_{cable}) and *Att.splitter.ant*$_a^i$ (the attenuation incurred by the splitters used to connect an antenna to the telephone room) into account. The former (attenuation of the cabling) is calculated as:

$$Att.cable.ant_a^i = \sum_{a \in A} dist_a \times \mathrm{att}_{cable} \times \tau_a^i \tag{2}$$

The calculation of the attenuation incurred by a sequence of splitters depends on the topology of the cabling, captured by the boolean $\kappa \in K$: $\kappa_{a,s}$ has the value 1 *iff* antenna a is connected by a cable to splitter s, and κ_{s_i,s_j} has the value 1 *iff* the the output of splitter s_i is connected to an input of splitter s_j. We use ρ_z^a to abbreviate connection sequences (e.g.: $\langle \kappa_{a,b}, \kappa_{b,c}, \ldots, \kappa_{x,y}, \kappa_{y,z} \rangle$) and defined it inductively:

$$\rho_{s_j}^{s_i} = 1 iff \begin{cases} \kappa_{s_i,s_j} = 1 \\ \exists \kappa_{s_i,s_k} : \kappa_{s_i,s_k} = 1 \text{ and } \rho_{s_j}^{s_k} = 1, \end{cases} \tag{3}$$

$$\rho_s^a = 1 \; iff \; \begin{cases} \kappa_{a,s} = 1 \\ \exists \kappa_{a,s_i} : \kappa_{a,s_i} = 1 \text{ and } \rho_s^{s_i} = 1, \end{cases} \tag{4}$$

The attenuation incurred by the connection through a sequence of splitters is:

$$Att.splitter.ant_a^i = \sum_{\substack{t \in \{2,3,4,6\} \\ s \in S_t}} \mathrm{att}_{splitter_t} \times \tau_s^i \times \rho_s^a \tag{5}$$

With $\mathrm{att}_{splitter_t}$ a constant for all splitters of type t (as discussed above).

From the above we can calculate the maximum attenuation incurred in floor f_i:

$$Att.at.floor_i = \max_{a \in A}(Att.cable.ant_a^i + Att.splitter.ant_a^i) \tag{6}$$

In order to meet the power requirements for all antennas on a floor f_i (each with their own requirement $power_a$) the minimum power input to the floor has to be:

$$Min.power.floor_i = \max_{a \in A}(Att.cable.ant_a^i + Att.splitter.ant_a^i + power_a) \tag{7}$$

The operational cost incurred for floor f_i, $Cost_{OpEx}^{f_i}$, is *at least Min.power.floor$_i$*.

3.2.3 A Solution to the Intra-Floor I-DAS Problem

The Intra-Floor Optimization problem for a floor f_i can be defined as an assignment to all $\tau_a^i, \tau_s^i \in T$, $\kappa_{a,s}, \kappa_{s_i,s_j}, \in K$ (and thus $\rho_s^a \in P$) such that the following are met: (1) antennas and splitters are allocated to a single specific floor and (2) there is exactly one splitter serving as entry node to each floor with the topology connecting antennas and splitters on the floor a non-cyclic (tree) graph with (a) all splitters being branching nodes, (b) all antennas being leaf nodes and (c) the entry node splitter being the root of that floor's topology tree. In other words, a solution to the single floor optimization problem is the network topology for this floor ($topology_{f_i}$), i.e. the information about which devices are on this floor and how they are interconnected: $topology_{f_i} = \langle K, T \rangle$ The performance value of a solution for any floor f_i can be expressed in terms of $Cost_{CapEx}^{f_i}$ (financial cost) and $Cost_{OpEx}^{f_i}$ (power requirements).

3.3 Solving the Problem at the Building Level (Inter-Floor)

3.3.1 Inter-Floor Hardware Cost

The Inter-Floor optimization problem can be seen as a variation on the Intra-Floor optimization, except we are connecting entire floors (represented by the floors main splitter s_{f_i}) instead of, as before, individual antennas. As before with the antennas, entire floors have their individual power requirements (which we calculated as $Min.power.floor_i$, Eq. 7). These have to be met to guarantee that the antennas on this floor are operational as designed. In addition, as before we have to consider the power attenuation incurred by the used splitters and cabling when connecting the ground floor main splitter to all floors. We are, again, interested to minimize the equipment and cabling cost. The incurred cost for the cabling is subject to our optimization as the splitters in this problem are distributed over the telephone rooms of the individual floors (with at most one splitter per floor). Without loss of generality we can simplify matters by assuming that the distance between floors is a constant: any two adjacent floors can be connected using the same length $dist_{floor}$ of cable.

We use the of boolean $\kappa_{j,i}$, introduced above, to encode the actual connections between individual floors and calculate $Cost_{CapEx}$ - the cost for the equipment and the cabling for the Inter-Floor problem *and* for the hardware required for the Intra-Floor problem for all floors - as follows:

$$Cost_{cables} = \sum_{i,j \in \{1,...,n\}} |i - j| \times dist_{floor} \times cost_{cable} \times \kappa_{j,i} \qquad (8)$$

$$Cost_{splitters} = \sum_{t \in \{2,3,4,6\}} cost_{splitter_t} \times |S_t \cap S_F| \qquad (9)$$

The hardware cost for the Inter-Floor problem can be calculated as follows:

$$Cost'_{CapEx} = Cost_{cables} + Cost_{splitters} \qquad (10)$$

The hardware cost for the I-DAS of the entire building also considers the cost for the hardware on all floors (cf. Eq. 1):

$$Cost_{CapEx} = Cost_{cables} + Cost_{splitters} + \sum_{f_i \in F} Cost_{CapEx}^{f_i} \qquad (11)$$

3.3.2 Inter-Floor Operational Cost

Similar to Eqs. 2 and 5 we can then calculate $Att.cable.floor_i$ (the attenuation incurred in the cabling between the base station on floor f_0 to the entry node on floor f_i, through arbitrary number of other splitters) and $Att.splitter.floor_i$ (the attenuation incurred along said path but from the used splitters), respectively:

$$Att.cable.floor_i = \sum_{s_j, s_k \in S} \text{dist}_{floor} \times |j - k| \times \text{att}_{cable} \times \kappa_k, j \qquad (12)$$

$$Att.splitter.floor_{s_{f_i}} = \sum_{\substack{t \in \{2,3,4,6\} \\ s \in S_t}} \text{att}_{splitter_t} \times \rho_s^{s_{f_i}} \qquad (13)$$

We calculate the attenuation incurred in a connection from the main base station to the entry node of a floor (Eq. 14), the resulting power requirements (Eq. 15) for this connection (given the power requirements of that floor, cf. Eq. 6):

$$Att.to.floor_i = Att.cable.floor_i + Att.splitter.floor_{s_{f_i}} \qquad (14)$$

$$Att.floor_i = Att.to.floor_i + Att.at.floor_i \qquad (15)$$

Using Eq. 15 to calculate the attenuation incurred by connecting a f_i floor, and including $Cost_{OpEx}^{f_i}$ (the minimum power requirements for this floor, cf. Eq. 7), we can now calculate the minimum power requirements for the entire building as follows:

$$Min.power.building = \max_{i \in \{1, \ldots, |F|\}} (Att.floor_i + Min.power.floor_i) \qquad (16)$$

The operational cost for the entire building, $Cost_{OpEx}$, is *at least Min.power.floor_i*.

3.3.3 A Solution to the Inter-Floor I-DAS Problem

Analogous to Sect. 3.2.3, the Inter-Floor Optimization problem for an entire building is the assignment of $\tau_s^i \in T$ and $\kappa_{s_i, s_j}, \in K$ such that the following are met: (1) there is at most one splitter per floor and (2) all floors connect to a single splitter, all splitters connect to the main splitter in the basement (except itself) and there are no cycles. In other words, a solution to the I-DAS optimization problem for an entire building is its network topology: *topology*, i.e. the information about where all devices are placed - and which devices they are connected to, i.e.: $topology = \langle K, T \rangle$. The performance value of a solution for the entire building can be expressed in terms of $Cost_{CapEx}$ (financial cost) and $Cost_{OpEx}$ (minimum power requirements).

4   Approach

As discussed in the previous section, we treat the I-DAS problem as a combination of two separate problems, with the solution of the single floor problem being part of the input to the building wide I-DAS optimization (see Fig. 6, p. 104). The problems differ significantly from each other and we use two different approaches to solve them: see Sect. 4.2 for our approach to optimize the topology of a floor, and Sect. 4.3 for the optimization of the topology connecting all floors to the base station.

We are using graphs to represent solutions to our optimization problems and for the building wide optimization we use Particle Swam Optimization (PSO, cf. Sect. 4.3.1) to heuristically search through numerical representation of all possible solution graphs. Section 4.1 discusses tree-graphs (our model for a solution) and provides the algorithms to translate them into a numerical representation (Prüfer Code, cf. [4, 11]).

4.1   Representing Network Topology as Tree Structures

We will represent solutions to the I-DAS optimization problem as graphs, specifically, as non-cyclic finite branching graphs, commonly called *trees*. In order to apply computational optimization approaches to our problem we need to represent these trees in a numerical format. Recall that *Cayley's Theorem* [7, 8] states that there are exactly n^{n-2} different trees with n nodes (we used this in Sect. 2 where we discussed the complexity of the I-DAS problem). Let us briefly refer to the proof of this theorem as it provides us with the insights and - by extension - the algorithms to encode trees efficiently: without going into the details, the first combinatorial proof provided for this theorem was provided by Prüfer [18] in 1918 [17] using a mapping that represented trees with n nodes as strings of length $n - 2$. By showing that the set of strings of that length had n^{n-2} members, Püfer proved *Cayley's Theorem*.

Since any tree can be uniquely represented by its Prüfer code [13], all possible spanning trees of n nodes can be represented by all $(n - 2)$-element Prüfer codes constructed from the set $\{1, .., n\}$. We will therefore use Prüfer string encoding to represent the search space for the I-DAS problem. We argue that, although Prüfer code has been shown to be a suboptimal choice as encoding for some optimization problems [10], they do work well in our case [4] (cf. Sect. 5 on our results).

Algorithm 1 Encoding a DAS-tree to a Prüfer Code (cf. [16])

1: $L \leftarrow$ leafs of T
2: **for** $i \leftarrow 1$ to $(n - 2)$ do **do**:
3:     $v \leftarrow$ node removed from the head of L
4:     $PC[i] \leftarrow$ neighbour of v
5:     delete v from T
6:     **if** $deg(PC[i]) = 1$ **then**
7:         add $PC[i]$ to L

The key advantage of using Prüfer code is that it encodes - uniquely - all possible valid tree structures and maps them into $n-2$ dimensional space. Such $n-2$ dimensional Euclidean spaces are known to work well with swarm and evolutionary search algorithms, such as Particle Swarm Optimization (cf. Sect. 4.3.1), used by us to solve the entire I-DAS optimization problem. With our intention to use Prüfer code as a mapping in mind we require an algorithm that can perform the translation from trees to Prüfer code (and back) efficiently. Encoding and decoding to and from Prüfer code follows a simple linear algorithm, details of which can be found in [13].

Algorithm 2 Decoding a Prüfer Code to a DAS-tree (cf. [16])

1: $L \leftarrow$ nodes that do not appear in the Prüfer code PC
2: **for** $i \leftarrow 1$ to $(n-2)$ do **do**:
3:     $v \leftarrow$ node removed from the head of L
4:     add edge $\{v, PC[i]\}$ to T
5:     **if** i is the rightmost position of v in C **then**
6:         add v to L
7:     $v \leftarrow$ node removed from the head of L
8:     add edge $\{v, PC[n-2]\}$ to T

The complexity of the Algorithms 1 and 2 is $\Theta(n)$ (i.e. the algorithms run in linear time). Once a tree is represented in Prüfer Code, comparing two strings and assigning a *distance* value to them is a straight forward process.

4.2 Solving the Intra-Floor I-DAS

We first discuss our approach to solve the network topology for individual floors, where the locations of the antennas (and thus the cable length required to connect them to the telephone room) are fixed. Figure 3 illustrates the inputs and outputs.

Whenever possible, building operators restrict the maximum number of splitters in a row (between base station and an antenna) to 3. Following this initial bounding – and with the mounting power attenuation incurred by the extensive daisy-chaining of splitters in mind – we focus on minimizing the depth of the connection tree.

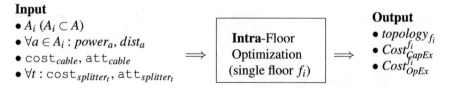

Input
- A_i ($A_i \subset A$)
- $\forall a \in A_i : power_a, dist_a$
- $cost_{cable}, att_{cable}$
- $\forall t : cost_{splitter_t}, att_{splitter_t}$

$\Longrightarrow$

Intra-Floor Optimization (single floor f_i)

$\Longrightarrow$

Output
- $topology_{f_i}$
- $Cost^{f_i}_{CapEx}$
- $Cost^{f_i}_{OpEx}$

Fig. 3 The Intra-Floor Optimization requires a (pre-optimized) layout for the antennas on a floor as well as hardware specifications and costs. The optimized topology can directly be used to calculate the resulting hardware cost as well as the operational power requirements for this floor

Initial (preliminary) investigations, considering variations of all 4 types of splitters (2-way, 3-way, 4-way and 6-way) suggested we focus our efforts on two extreme strategies, which emerged as the by far most promising ones:

1. **maximum peripheral branching**: the end nodes of the connection tree are connected using only the *largest* available splitter type (in our case this is the 6-way splitter). Any left-over antennas can be connected by using one different (i.e. a 4-, 3- or 2- way) splitter in the end.
2. **minimum peripheral branching**: the last level of the connection tree is binary, i.e. constructed entirely from the *smallest* available splitter type (i.e. the 2-way splitter). In the case of an odd number of antennas at any tree level, one of the splitters is replaced by a 3-way splitter to include the leftover antenna and thereby avoid unbalanced power distribution or power loss with attenuators.

In both cases the remaining levels of the connection tree are constructed according to straight forward heuristics and the assignment of antennas to splitter ports is subjected to the constraint that the antenna with the highest power requirement must be connected to the splitter with the lowest accumulative power attenuation. We then use a guided process to locally improve the so generated solution and to furthermore ensure that the excess power delivered to the individual antennas does not exceed 2dbm. The latter may also require replacing 6-way splitters with splitters of lower connectivity (for this, as mentioned above, we biased our choice towards 2-way splitters). This - while not guaranteeing the lowest possible power requirements for a floor - allows us to to quickly generate good (albeit not necessarily optimal) solutions to the Intra-Floor DAS Problem. Cf. Figs. 2 and 4 for illustrations.

The tree is constructed such that the maximum depth is minimized, i.e. splitters connect only to splitters until the combined available outputs of the connected splitters are equal or greater than the number of antennas on the floor (i.e. $\leq |A|$).

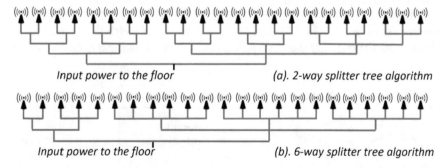

Input power to the floor   *(a). 2-way splitter tree algorithm*

Input power to the floor   *(b). 6-way splitter tree algorithm*

Fig. 4 A visual representation of the 2- and 6-way splitter tree algorithm for 23 antennas. The blue, green and pink tree branches correspond to the 2-, 3- and 6- way splitters respectively. The order of connections follows the logic of connecting first the antennas that require more power to the ports of splitters with smaller power loss

Table 2 A cost comparison between 2-, 3-, 4- and 6-way splitter algorithm applied to 15 antennas (2015 US$)

Splitter type	2-way	3-way	4-way	6-way
Cost (USD)	694	556	459	378
Power attenuation (dBm)	13.8	12.6	12	12.6

Using the cost and attenuation for the splitter types given in Table 2, the 6-way splitting algorithm outperforms the 2-way splitting algorithm in both, minimizing the cost ($Cost^i_{CapEx}$, Eq. 1) as well as insuring a low attenuation ($Att.floor_i$, Eq. 15).

As a general comment (adding to the above): it is better to start out using the largest possible splitters, because the 6-way algorithm maximally reduces the depth of the tree. When power consumption is factored in, the 2-way algorithm typically results in more splitters along some paths, translating into higher power loss.

4.3 Solving the Inter-Floor I-DAS

Figure 5 illustrates the inputs to the Inter-Floor optimization approach. The topologies (and resulting equipment cost and incurred power attenuation) used are the output of the optimization approach discussed in the previous section.

In contrast to the intra-floor optimization, for the inter-floor optimization we know the size of the resulting tree (i.e. we know the number of nodes the tree will have). This is due to a simple consideration: when optimizing a single floor we were given the set of antennas (the *leaf nodes* of the tree) and we added splitters to the topology (the *branching nodes*) until all antennas were connected to a single splitter (the *root*) which served as the entry node to the floor. The resulting tree was therefore of the size *leaf nodes* + *branching nodes*. However, when optimizing the entire building we use the roots of the individual floors as *leaf nodes*, of which there are as many as there are floors. Other than for the individual floor, the *branching nodes* are not

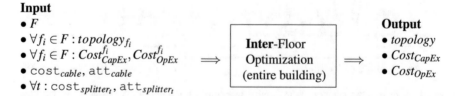

Fig. 5 Input and Output of the Inter-Floor optimization approach we used. The approach uses Particle Swarm Optimization to heuristically explore a $n - 2$ dimensional Euclidean solution space (with n the number of nodes in the I-DAS topology), i.e. the Prüfer Code for the n^{n-2} different trees (I-DAS topologies) and returns a single preferred topology as well as the resulting hardware cost (CapEx) and the building's minimum power requirements (OpEx)

required to be in a central location (i.e. the location of the I-DAS base station in the basement of the building), and it is straight forward to see that placing the splitters next to the entry nodes of the floors is beneficial as it reduces the overall amount of cable required. When comparing this to the floor level optimization, this would be analogous to placing the splitters co-located with individual antennas instead of in the floor's telephone room. On the floor level this is impractical as the splitters are best co-located in one secure place, but for the building level optimization the same reasoning applies: the splitters can be located in the telephone rooms of the floors (placing them next to the floor's entry node) and thus be distributed throughout the building. Since placing splitters with the antennas reduces the cabling it follows that no two splitters should be placed next to the same antenna, thus there is at most one splitter on any floor. This means that the nodes of our topology are either just an entry node (a *leaf node*, no splitter) or an entry node connected directly to a splitter (a *branching node* that is also a *leaf node*). Therefore (since all branching nodes are leaf nodes) the number of nodes in our tree is equal to the number of floors.

Thus, the advantage of distributing the splitters throughout the building is that - for the inter-floor optimization - the size of the topology is known in advance. The disadvantage, however, is that we can no longer ignore the cabling required to connect the splitters to one another (on a floor level this was negligible as splitters are stacked in the same room) as this can be as long as the largest distance between any two floors (and thus substantial). Therefore, the simple heuristic discussed in Sect. 4.2 would not work for the optimization of I-DAS topology for the entire building. Instead we apply swarm intelligence in form of Particle Swarm Optimization (PSO) to heuristically explore the solution space (which, since we know n, the number of nodes in the tree, can be encoded as the set of Prüfer Strings of size $n - 2$; cf. Sect. 4.1).

4.3.1 Particle Swarm Optimization (PSO)

Particle Swarm Optimization (PSO) is a meta-heuristic where particles embodying the candidate solutions move through the solution space (in various directions) at different velocities, thereby imitating natural swarms of birds, insects or fish [23].

Consider this simple example to illustrate how and why PSD works: a school of fish is swimming in a lake looking for food. None of the fish has any knowledge about where the food is, but all of the fish can see the direction and the speed of their immediate neighbours. By continuously adapting their own orientation and velocity to that of their neighbours, all members of the school become part of the school and - using only this information - the entire swarm moves in a seemingly coordinated way. In nature this turns into a simple *follow the leader* game, where the leader is determined by its position in the swarm. PSO leverages the mechanisms used by the members of the school to achieve collective coordination in the following way: if, hypothetically, the leader of the school was determined by some process that took the distance to the food into account, then we'd have a swarm that always followed that member which is the closest to the food (i.e. the optimal location). While we do not know the location of the food (the optimal solution) we can determine which fish is closest to it by simply comparing their locations (i.e. the values of their solutions).

4.3.2 Algorithm

In PSO, a swarm of individual particles explores the solution space in order to find the best locations, corresponding to the best solution. To achieve this goal all particles initially *fly* along random directions and with random velocities. Gradually, through interaction with other particles their exploration directions are adjusted and eventually the swarm converges on an area in the solution space that is likely to contain the best or very good solutions. The increasingly coordinated movement is the key feature of PSO and it distinguishes it from other evolutionary algorithms. In [14] performance improvement of PSO was achieved by adding randomness to the velocity to avoid the traps of local optima.

Algorithm 3 PSO-DAS model pseudocode

1: **function** (P, C) = PSO- DAS(PARTICLEN, FLOORN)
2:   **for** each particle i **do**:
3:       Init random position $x_i \in \{1, .., n\}$ and random velocity $v_i \in \{v_{min}, .., v_{max}\}$,
4:       Init best position and cost: $p_i \leftarrow x_i$; $c_i \leftarrow \infty$ (local) and $P \leftarrow x_i$; $C \leftarrow \infty$ (global)
5:   **while** not (termination criterion) **do**
6:       **for** each particle i **do**:
7:           **if** x_i passes power constraints **then**
8:               Evaluate cost function $C(x_i)$
9:               **if** $C(x_i) < c_i$ **then** $p_i \leftarrow x_i, c_i \leftarrow C(x_i)$
10:                  **if** $c_i < C$ **then** $C \leftarrow c_i, P \leftarrow p_i$
11:              **for** each dimension $d = 1, .., n - 2$ **do**
12:                  $v_{i,d} = \alpha v_{i,d} + \beta r_p(p_{i,d} - x_{i,d}) + \gamma r_g(g_{i,d} - x_{i,d})$
13:              Update particle's position $x_i \leftarrow x_i + v_i$

In optimization techniques such as PSO we map all possible solutions onto some domain and then *move* through that domain in a somewhat coordinated fashion. It is often straight forward to calculate the cost (commonly called *fitness*) of a specific solution (i.e. the solution that is represented by the location I am currently in), determining the member of the swarm that currently has the best solution is a straight forward process. Moreover so, determining the new leader of the swarm comes at a computational cost that is linear in the size of the swarm (i.e. if we have done our job right and picked a computationally efficient encoding algorithms, this should not cause us any headaches). Orientation and speed, with respect to moving through the solution space, translate into the direction of a vector and the length. For any particle, given these two values and its current position, we can calculate the location where it will be next. A collection of computational agents (called particles) can thus be initialized randomly and let loose on a solution space. They will flock to their leader, who is determined, each turn, on the basis of the best found solution. This results in guided exploration of the solution space and convergence on good solutions. Obviously, the parameters that control the swarming and the speed of convergence are crucial in avoiding either searching too long or converging too quickly (thereby zooming right past the best solution without noticing it).

4.4 Multiple Optimization Objectives

The above has, so far, only included the hardware cost (CapEx) while the operational power consumption of the system (OpEx) is considered exclusively in the single-floor level optimization where it was factored into the solutions only insofar as we require antennas to receive no more than 2 dBm *more* than they require to operate.

4.4.1 Floor Level Hardware Cost/Power Requirements Trade-Off

The power demands of the antennas affect the power requirements of the individual floors (i.e., the leaf nodes of the inter-floor topology). While the operational parameters of the antennas themselves are fixed, the resulting power requirements for the individual floors can often be decreased (by optimizing the intra-floor problem). This may result in a significant reduction of the building-wide power requirements at the price of increased hardware cost on some floors. There is clearly a trade-off and, with it, a potential advantage in deploying otherwise sub-optimal hardware configurations for some floors (when, e.g. specific floors prevent an extremely good solution from being feasible for the entire building). If we amend the floor level optimization (Sect. 4.2) to consider an acceptable spread for the hardware cost (instead of optimizing for it) we may improve on the floor's power requirements and, since these are the inputs to the building optimization (Sect. 4.3), thereby achieve a better overall solution.

4.4.2 Building Level Hardware Cost/Power Requirements Trade-Off

If we want to consider optimizing the power requirements of the entire building at the inter-floor level we have to amend the fitness function of the building optimization. Section 4.3 the PSO is said to evaluate the value of a candidate solution but so far we have suggested considering the aggregated hardware cost as the performance value for solutions. So far the cost C (cf. Algorithm 3) is simply:

$$C = Cost_{CapEx}$$

Let us amend this to include the excess power delivered to the antennas:

$$C' = (1 - \eta)Cost_{CapEx} + \eta Power_{dev} \tag{17}$$

And distinguish $Power_{dev}$ further:

$$Power_{dev} = (1 - \varepsilon)Power_{\sum dev} + \varepsilon Power_{max\_dev} \qquad (18)$$

with η and ε tuning parameters **with values > 0 and < 1**, to adjust the weighting and $Power_{max\_dev}$ the largest excess power (positive deviation from the required power) of any antenna, and, similarly, $Power_{\sum dev}$ the aggregated excess power of all antennas. Using this amended cost function our approach can be tuned to consider both, $Cost_{CapEx}$ and $Cost_{OpEx}$.

4.5 Implementation

The implementation of the approach has to bring together the various parts and aspects discussed above. The PSO was applied to the inter-floor problem, generating new generations of suitable network topologies to connect the different floors. The evaluation of these solutions included varying maximum power budgets and – of course – relied heavily on the power requirements of the individual floors (i.e., relied on the current solution for the intra-floor problem). Figure 6 shows an overview over the information required by the system and how it flows.

As shown in Fig. 7, the PSO continuously loops between generating a set of new candidate solutions and passing these candidate solutions to an evaluation module,

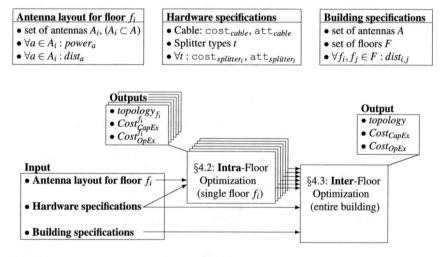

Fig. 6 An overview over the information flow in our approach. The single floor optimization, discussed in Sect. 4.2, provides part of the input to the optimization for the entire building (cf. Sect. 4.3)

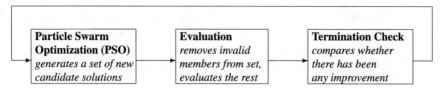

Fig. 7 The high-level view on the approach: a PSD is generating candidate solutions which are checked against basic validity criteria before being assigned a performance value (the overall cost for the solution). On the basis of the so vetted and evaluated solutions, a new set of solutions is generated by the PSO. The process continues until some termination condition is met

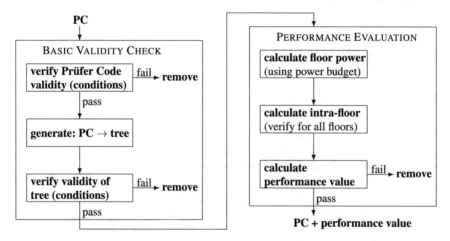

Fig. 8 The stages of the evaluation process. These are applied to every individual candidate solution generated by the PSO: first the basic validity of the solution is verified by ensuring that the solution does meet minimum requirements such as starting in the root node and using only available hardware. In the subsequent performance evaluation the maximum available power to each floor is calculated (using a variable power budget) and solutions that leave any floor underpowered are removed. Finally the cost C' is calculated and returned together with the solution

where a fitness value is assigned to each one. Solutions are then chose as leader for the swarm for the next iteration by comparing their fitness values.

Clearly, the evaluation module as well as the stop condition are crucial for the implementation. The evaluation module, visualized in Fig. 8, works as follows:

1. A set of candidate solutions (each written as Püfer Code) is received
2. The set of PCs is subjected to straight forward validity criteria. E.g., any solution having a branching factor above 6 or the wrong root is removed from the set.
3. The remaining set of PCs is decoded – individually – into the tree representation, which is then subjected to the following validity criteria:

   a. Basic properties of the tree such as e.g., a maximum depth of 3, are checked. Offenders are rejected from the set.

b. Considering the current power-budget settings (at the root) the resulting power at the leafs is calculated (bottom up). Intra-floor solution for each floor are checked against these floor specific power-budgets. If any floors requires more power than the leaf can provide, the solution is rejected and removed.

4. The remaining set now only contains topologies that match the basic requirement for branching and depth, as well as are known to be within the maximum power budgets. Using Eq. 17 for our revised cost function we can calculate a performance value for each remaining member of the candidate solution set.

5 Performance Evaluation and Results

5.1 Algorithm Performance

Evaluating the approach against the global optimum (i.e. comparing the best found solution to the best possible solution) was hampered by the $\Theta(n^{n-2})$ complexity of the problem. For buildings with up to $n = 8$ floors the best possible solution was determined through exhaustive search and in these cases the PSO-DAS approach also found the same optimal solution. To mention computation times for comparison: for the $n = 8$ problem the exhaustive search took 15 min while our approach returned the optimal solution after 15 s. Given that we project an exhaustive search for $n = 13$ to take around 75 years we settle for the claim that the approach vastly outperformed the exhaustive search with regard to time while producing optimal solutions for $n \leq 8$ and very good or near optimal solutions for larger n.

5.1.1 PSO Model Internal Assessment

To investigate the PSO progress during execution we compared the top solution found in the initial random population against the one found in the last generation.

Figure 9 shows a comparison for up to 100 floors and the typical 50–70% improvement in cost reduction compared to the initial random best solution. Somewhat higher relative improvement at lower floor numbers is understandable because of the increased influence of the power budget constraint for the very tall buildings.

5.1.2 PSO, GA and Using a Hybrid Approach

In this chapter we primarily emphasise our PSO based approach. In Chap. 5 we briefly discuss other heuristics and of course the modelled problem can be subjected to other approaches such as genetic algorithms (GA). To validate the performance of our PSO implementation we also developed a GA-based implementation and compared their

Fig. 9 Comparison of the best solution (least cost) drawn from the last vs the first PSO generation

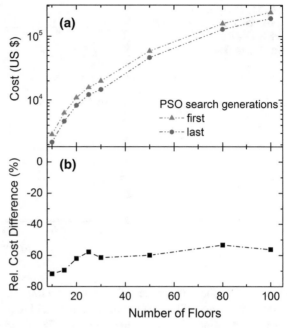

Fig. 10 Comparison: the incurred computational cost of the PSO and the GA implementation

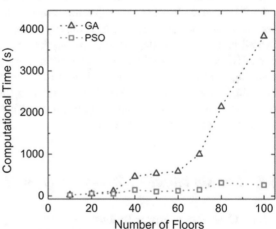

performances. Figure 10 shows the difference in computational time between a PSO-only and a GA-only implementation: the PSO consistently outperforms the GA for increasing problem sizes.

Figure 11 compares the relative (i.e., PSO over GA) improvement with regard to the cost of the generated solution. Again, PSO outperforms GA significantly.

In addition a hybrid implementation, combining PSO and GA elements (as opposed to PSO-only), was developed. Figure 12 shows that the hybrid approach

Fig. 11 Comparison: the projected deployment cost of the PSO and the GA implementation

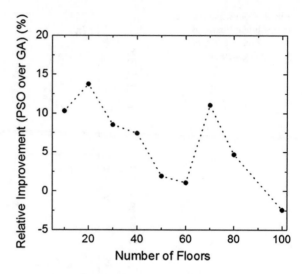

improved on the performance of the PSO-only implementation, leveraging the superior performance of the PSO implementation and improving on it using subsequent GA optimization of the generated solutions. The hybrid approach outperformed the *all-PSO* approach in all cases, delivering lower deployment cost and requiring less computational power to do so. In a hindsight, the *all-PSO* approach hinders the overall performance without producing any benefits for the cost optimization outcome. This is so because the all-PSO algorithm explores a much larger solution space resulting in the longer computing time. Realistically speaking, an I-DAS deployment on a single floor is unlikely to exceed 20 antennas (usually 8–20, in most practical realizations) and for these small problems our heuristic worked well, motivating the hybrid approach proposed and tested.

5.2 Cost Optimization Versus Power Budget Constraints

5.2.1 Cost Versus Power Budget

As our stated goal is to address excessive power consumption and to design sustainable infrastructures, we imposed power budget restrictions on the solutions (i.e. we artificially restricted the maximum power available at the base station), thereby incentivizing the finding of solutions with lower power requirements (at the price of a higher initial hardware cost).

Figure 13 illustrates the evolution of minimum-cost solutions found by PSO-DAS for buildings with a rising number of floors in the following power budget versions: 40, 45, 50 and 60 dBm. As can be seen, minimizing the power consumption and the cabling cost are two opposing objectives and a tradeoff between the two is needed

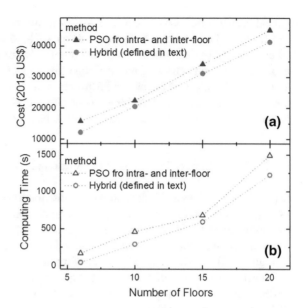

Fig. 12 Comparison: a PSO-only versus the hybrid PSO+GA implementation (also using GA)

to achieve a good balanced solution. Limiting the power budget forces the model to return worse-cost solutions and vice versa. The limitations of this are illustrated by the fact that increasing the power budget past a certain limit does no longer return less expensive solutions.

The balance between minimizing the power budget or the deployment cost can be controlled depending on the priorities of the building management. As we argue in the introduction, adopting a resource aware stance is imperative, however, in a

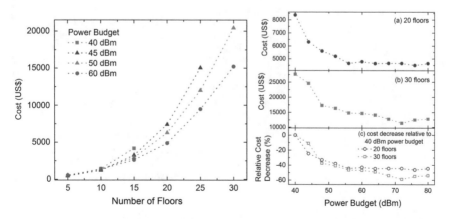

Fig. 13 The effect of different power constraints on cost with the last plotted point being the terminating data point (left) and the relationship between power budget and inter-floor deployment cost for the same size 20-floors building (right)

typical scenario the power budget is assumed upfront and set as a constraint for cost minimization. In our subsequent analysis, we have increased power budgets for taller buildings in line with the empirical observation of making the PSO search for good cost-solutions challenging. We hope that - given this efficient tool - I-DAS design decision can now be made with an eye on the running costs of the installation; translating sustainable I-DAS deployment into long-term financial benefit has the potential to make this part of the IoT infrastructure less power consuming.

5.2.2 Power Budget Versus Feasible Solutions

In practice, variations and imbalances in the distribution of the antennas are the cause for large differences in the resulting power requirements. This issue can be addressed by deploying power attenuators to improve the effect of power-management aware solution design (in fact, this is common practice by practitioners in the field).

Figure 14 shows the returned number of feasible solutions (for a specific example problem) as a function of the power-budget constraint. The evaluated range (35–60 dB) is informed by the practice in the field. In this ministudy, the number of floors in the building was considered a controllable parameter.

Fig. 14 Panel **a** shows the output with an additional 2dB power-attenuation criterion, while panel **b** shows the situation without power attenuators being included

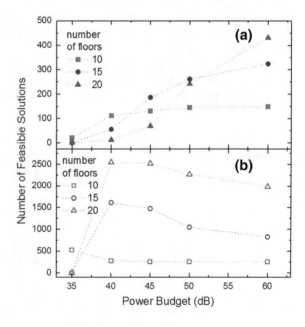

Fig. 15 The impact of increasing the power-budget: more, as well as consistently cheaper solutions are produced as the power-budget constraints are relaxed

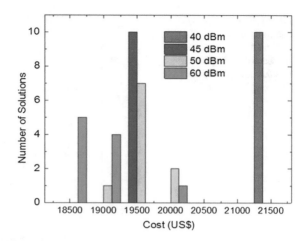

5.2.3 Power Budget Versus Solution Diversity and Reliability

The diversity of the returned solutions (remember, multiple solutions were returned to offer additional flexibility to the practitioner in the field) was evaluated in the context of the (unique) number of returned solutions (grouped by their projected cost). Figure 15 represents a study of diversity and reliability of the generated solutions for a specific example problem. Already for the tight constraints of 40 and 45 dBm there is a clear trend in that increasing the overall power-budget results in a decreased (projected) cost. For more generous budgets (here 50 and 60 dBm), this also lead to a diversification of the solutions in that there was an increasing spread across the projected costs as well. The used example problem was a 20-floor building – similar studies, performed for other problem sizes (number of floors) produced qualitatively equivalent outcomes.

5.3 Practical Considerations

5.3.1 PSO-DAS Termination Criterion

One of our initial investigations was designed to gain insight into what constitutes a suitable termination criterion for Algorithm 3. To this end 30 runs were conducted, applying the PSO-DAS model to problems with n ranging from $n = 5, \ldots, 100$. We recorded the probability of improving upon the current best found solution in dependance to the number of previous consecutive iterations without improvement. The results are plotted in Fig. 16. We found that after 10 consecutive generations without improvement the probability of improving further approaches zero.

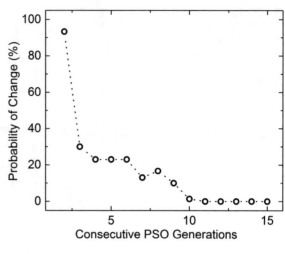

Fig. 16 The termination criterion for the PSO is a minimum number of consecutive PSO generations without an improved solution

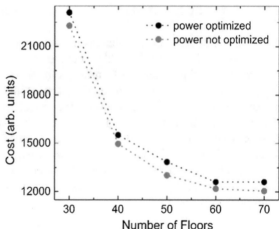

Fig. 17 The (average) hardware cost (i.e., the one time investment cost of implementing the solution) for the final solutions when the solution was, or was not, also optimized for power requirements (i.e., the continuously incurred operational cost of the solution)

5.3.2 Deployment Cost Versus Operational Cost

Early stages of the research in this study were conducted with the goal of satisfying the criterion of minimized cost of deployment, as this is one of practical considerations driving the project. However, the goal of optimizing power in I-DAS system is an issue of comparable importance. In Fig. 17, we plot the cost for hardware deployment of solutions generated for cost-reduction only, as well as for a combined cost-reduction/operational-cost based fitness-function (cf. Eqs. 17 and 18). The power optimization criterion results in increased deployment cost, however, as the operational costs are incurred continuously smaller increases will be offset by the reduced operational cost over time.

Optimizing the total deployment cost of DAS is only one objective in the overall task of network optimization. Another, equally important task is optimization of the signal power delivered to each antenna within the valid range. To consider this problem, we again rely on realistic scenarios, where a power deviation constraint is provided to reduce the cases of which antennas receive an excess of power. It is clear from our discussion that optimizing the power deviation among all the antennas clashes with the objective of optimizing the total deployment cost of the system, because minimizing the power deviation results in returning a solution with high deployment cost, and vice versa. The requirement of simultaneous satisfaction of both objectives leads to a multi-objective optimization function.

Recall Eqs. 17 and 18 from p. 17, defining our amended cost function:

$$C' = (1 - \eta)Cost_{CapEx} + \eta Power_{dev}$$

$$Power_{dev} = (1 - \varepsilon)Power_{\sum dev} + \varepsilon Power_{max\_dev}$$

The coefficients η and $(1 - \eta)$ are the weight factors given for both optimization variables where the summation of both coefficients is equal to 1 and same applies to ε and $(1 - \varepsilon)$. The weight given to each variable reflects the extent of focus that will be given to that variable in the optimization process. In order to find the optimal balance between both optimization variables, both variables should be assigned equal weights before the optimization is performed, where $\eta = \varepsilon = 0.5$. Empirical studies were been conducted to choose values for η and ε.

Tables 3 and 4 summarise the results of two numerical experiments showing that the best values for η and ε which return the best solution (highlighted in the two tables) are 0.9 and 0.7, respectively. These empirically driven values were the result of four numerical experiments examining the effect of different values of η and ε on the total deployment cost, maximum power deviation, power deviation summation and the final result of the fitness function. Except for power deviation summation test (favoring $\eta = 0.5$ and $\varepsilon = 0.9$) all results converged to the same conclusion.

Table 3 Parameter-space exploration for the parameters η and ε with regard to the hardware cost of the solutions. The highlight values clearly indicate the optimal parameter values

$\varepsilon \backslash \eta$	0.1	0.3	0.5	0.7	0.9
0.1	15,135.57	14,043.33	13,073.22	14,454.72	14,454.72
0.3	15,140.97	15,568.56	14,054.13	14,327.19	15,703.29
0.5	13,331.88	14,449.32	14,603.85	13,353.48	15,976.35
0.7	15,008.04	14,050.53	13,768.47	13,764.87	**12,388.77**
0.9	14,871.51	14,179.86	14,727.78	14,174.46	14,048.73

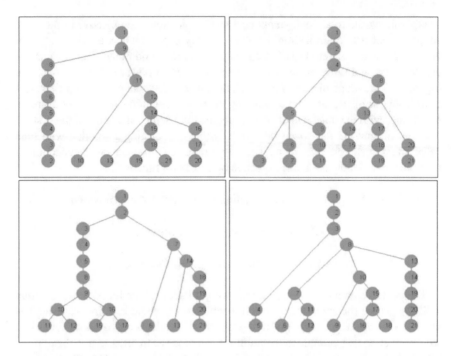

Fig. 18 Four different solution trees for one unique problem (building of 21 floors, with the 42 dBm power constraint), optimized for cost, i.e., without simultaneous power-cost evaluation (top left, right) and optimized for power as well as cost (bottom left, right)

5.3.3 Generating Multiple Candidate Solutions

The system outputs a number of solutions (as opposed to just the one with the best found performance), e.g., Fig. 18. This enables the practitioner to be more flexible during the implementation phase and pick the solution that does fit best, given that every building is different and the individual decision maker may have personal preferences not captured in our formal model.

Table 4 Parameter-space exploration for the parameters η and ε with regard to the power requirements of the solutions. The highlight values clearly indicate the optimal parameter values

$\varepsilon \backslash \eta$	0.1	0.3	0.5	0.7	0.9
0.1	11.62	11.58	19.25	12.08	12.05
0.3	12.18	13.25	11.24	14.45	12.65
0.5	11.62	11.58	13.25	13.38	11.58
0.7	10.39	19.25	14.45	13.38	**9.55**
0.9	13.25	14.44	16.24	13.38	15.08

6 Conclusions

We provide a modelling for wireless access network topologies, specifically, for Indoor Distributed Antenna Systems (I-DAS). These networks of wireless access points often constitute the *last mile* of the global internet and are widely used in today's increasingly connected world. The difference in hardware cost, as well as operational requirements (here: power) can differ greatly between topology variations but due to the complexity of these networks, using traditional exhaustive approaches to determine an optimal network topology quickly becomes unfeasible.

Distributed, nature-inspired approaches have been applied to a wide variety of computationally intractable problems and have been shown to reliably produce good (albeit not optimal) solutions. Such *heuristics* are scalable and robust, motivating their use for classes of problems such as the I-DAS design problem.

In this chapter, we have provided a modelling for I-DAS network topologies and presented results from a hybrid Particle Swarm Optimization (PSO)/Genetic Algorithm (GA) approach applied to large I-DAS optimization problems. The chapter provides details of our modelling as well as the algorithms required to reproduce our work. Design choices are motivated and common practices for the design of I-DAS, where we are aware of them, are described.

Within the limits of exhaustive search (where determining the best solution is computationally feasible), the presented approach finds the best possible solutions; the results obtained when applying the system to (much) larger problems match the extrapolated projections, indicating a very good performance of the approach. In any case, the results generated for problems of up to 100 floors are feasible from a practical and operational point of view, making the approach useable in the field.

With regard to the use of PSO, we have also addressed the issue of encoding the solution space into n-dimensional Prüfer Space in Chap. 5, in this chapter we simply provide the algorithms and relevant complexity results.

Finally, it has not escaped our attention that the approach, while detailed to address I-DAS problems, has the potential to be applied to similar problems in the domain of sustainable computing and infrastructure deployment and design.

Acknowledgements The authors are grateful for the support from the UAE ICT-Fund on the project *"Biologically Inspired Network Services"*. We acknowledge K. Poon (EBTIC, KUST) for bringing the I-DAS problem to our attention. HH acknowledges the hospitality of the EBTIC Institute and F. Saffre (EBTIC, KUST) during his fellowship 2017.

References

1. Adjiashvili, D., Bosio, S., Li, Y., Yuan, D.: Exact and approximation algorithms for optimal equipment selection in deploying in-building distributed antenna systems. IEEE Trans. Mob. Comput. **14**(4), 702–713 (2015)

2. Atawia, R., Ashour, M., El Shabrawy, T., Hammad, H.: Indoor distributed antenna system planning with optimized antenna power using genetic algorithm. In: 2013 IEEE 78th Vehicular Technology Conference (VTC Fall), pp. 1–6 (September 2013)
3. Atia, D.Y.: Indoor distributed antenna systems deployment optimization with particle swarm optimization. M.Sc. thesis, Khalifa University of Science, Technology and Research (2015)
4. Atia, D.Y., Ruta, D., Poon, K., Ouali, A., Isakovic, A.F.: Cost effective, scalable design of indoor distributed antenna systems based on particle swarm optimization and Prüfer strings. In: IEEE Proceedings of 2016 IEEE Congress on Evolutionary Computation, Vancouver, Canada (July 2016)
5. Beijner, H.: The importance of in-building solutions in third-generation networks. Ericson Rev. **2**, 90 (2004)
6. Borchardt, C.W.: über eine Interpolationsformel für eine Art symmetrischer Funktionen und über deren Anwendung. In: Math. Abh. Akad. Wiss. zu Berlin, pp. 1–20. Berlin (1860)
7. Cayley, A.: On the theory of the analytical forms called trees. Phil. Mag. **13**, 172–6 (1857)
8. Cayley, A.: Volume 13 of Cambridge Library Collection - Mathematics, pp. 26–28. Cambridge University Press, Cambridge (July 2009)
9. Chen, L., Yuan, D.: Mathematical modeling for optimal design of in-building distributed antenna systems. Comput. Netw. **57**(17), 3428–3445 (2013)
10. Gottlieb, J., Julstrom, B.A., Raidl, G.R., Rothlauf, F.: Prüfer numbers: a poor representation of spanning trees for evolutionary search. In: Proceedings of the Genetic and Evolutionary Computation Conference (GECCO 2001), pp. 343–350, San Francisco, CA, USA. Morgan Kaufmann Publishers, Burlington (2001)
11. Hildmann, H., Rta, D., Atia, D.Y., Isakovic, A.F.: Using branching-property preserving Prüfer code to encode solutions for particle swarm optimisation. In: 2017 Federated Conference on Computer Science and Information Systems (FedCSIS), pp. 429–432 (September 2017)
12. Hiltunen, K., Olin, B., Lundevall, M.: Using dedicated in-building systems to improve HSDPA indoor coverage and capacity. In: 61st IEEE Conference on Vehicular Technology, pp. 2379–2383 (2005)
13. Julstrom, B.A.: Quick decoding and encoding of Prüfer strings: exercises in data structures (2005)
14. Ma, R.-J., Yu, N.-Y., Hu, J.-Y.: Application of particle swarm optimization algorithm in the heating system planning problem. Sci. World J. (2013)
15. Marzetta, T.L.: Noncooperative cellular wireless with unlimited numbers of base station antennas. IEEE Trans. Wireless Commun. **9**(11), 3590–3600 (2010)
16. Micikevičius, P., Caminiti, S., Deo, N.: Linear-time algorithms for encoding trees as sequences of node labels (2007)
17. Paulden, T., Smith, D.K.: Developing new locality results for the Prüfer code using a remarkable linear-time decoding algorithm. Electr. J. Comb. **14** (2007)
18. Prüfer, H.: Neuer Beweis eines Satzes über Permutationen. Archiv der Mathematik und Physik **27**, 742–744 (1918)
19. Ren, H., Liu, N., Pan, C., He, C.: Energy efficiency optimization for MIMO distributed antenna systems. IEEE Trans. Veh. Technol. **99**, 1–1 (2016)
20. Sun, Q., Jin, S., Wang, J., Zhang, Y., Gao, X., Wong, K.K.: On scheduling for massive distributed MIMO downlink. In: 2013 IEEE Global Communications Conference (GLOBECOM), pp. 4151–4156 (December 2013)
21. You, X.H., Wang, D.M., Sheng, B., Gao, X.Q., Zhao, X.S., Chen, M.: Cooperative distributed antenna systems for mobile communications [coordinated and distributed MIMO]. IEEE Wirel. Commun. **17**(3), 35–43 (2010)
22. Ross, Y., Watteyne, T.: Reliable, low power wireless sensor networks for the internet of things: making wireless sensors as accessible as web servers. White paper, Linear Technology (December 2013)
23. Zhou, L., Li, B., Wang, F.: Particle swarm optimization model of distributed network planning. JNW **8**(10), 2263–2268 (2013)

InterCriteria Analysis Approach for Comparison of Simple and Multi-population Genetic Algorithms Performance

Maria Angelova and Tania Pencheva

Abstract Intercriteria analysis approach, based on the apparatuses of index matrices and intuitionistic fuzzy sets, has been here applied for comparison of simple and multi-population genetic algorithms performance. Six kinds of simple genetic algorithms and six kinds of multi-population genetic algorithms, differing in the execution order of the main genetic operators selection, crossover and mutation are in the focus of current investigation. Intercriteria analysis approach is implemented to assess the performance of mentioned above genetic algorithms for the purposes of parameter identification of *Saccharomyces cerevisiae* fed-batch fermentation process. For the completeness, the performance of altogether twelve algorithms have been also assessed. Degrees of agreement and disagreement between the algorithms outcomes, namely convergence time and model accuracy, from one hand, and model parameters estimations, from the other hand, have been established. The obtained results after the application of intercriteria analysis have been compared and outlined relations have been thoroughly discussed.

1 Introduction

Yeast are widely used model eukaryotic organisms in contemporary molecular and cell biology due to its well known metabolic pathways [1, 18]. *S. cerevisiae* has been intensively to become studying for the production of medicines, foods and beverages in numerous investigations. Also, it is the microorganism behind the most common type of fermentation. However, fermentation processes are well known with the complex structure of their mathematical models, usually described by systems of nonlinear differential equations and a number of specific growth rates. In gen-

M. Angelova (✉) · T. Pencheva
Institute of Biophysics and Biomedical Engineering,
Bulgarian Academy of Sciences, Sofia, Bulgaria
e-mail: maria.angelova@biomed.bas.bg

T. Pencheva
e-mail: tania.pencheva@biomed.bas.bg

© Springer Nature Switzerland AG 2019
S. Fidanova (ed.), *Recent Advances in Computational Optimization*,
Studies in Computational Intelligence 795,
https://doi.org/10.1007/978-3-319-99648-6_7

eral, modelling of fermentation processes is a real challenge for investigators. Model parameter identification is of a key importance for solving of such a difficult task. Genetic algorithm (GA), as representatives of biologically inspired metaheuristic techniques, have pointed more and more attention mainly due to the fact that they reach a good solution in the field of complex dynamic systems optimization [13, 14, 19], among them for parameter identification of various fermentation process models [1, 2, 17–19]. GA are one of the methods based on Darwin's theory of survival of the fittest. Simple genetic algorithms (SGA), originally presented in [14], search a global optimal solution using three main genetic operators in a sequence selection, crossover and mutation over the individuals in one population. Meanwhile, multi-population genetic algorithms (MpGA) is more similar to the nature since in them many populations, called subpopulations, evolve independently from each other. After a certain number of generations, a part of individuals migrates between the subpopulations.

Any kind of detected relations between model parameters and optimization algorithm performance might lead to significant improvement of the parameter identification procedure. InterCriteria Analysis (ICrA) [12] is going to be applied aiming to achieve such improvement. ICrA is an approach based on the fundamental concepts of intuitionistic fuzzy sets [6–8] and index matrices [9–11]. ICrA allows finding possible correlations between pairs of involved criteria and provides on this basis an additional information for the investigated objects. So far the approach of ICrA has been successfully applied in different fields of science, including medicine [21, 22]. Successful implementation of ICrA for assessing the performance of six simple genetic algorithms for parameter identification of a *S. cerevisiae* fermentation process [16] born the idea the approach to be implemented for the same purpose, but applying six multi-population genetic algorithms [3]. Outlined relations in [16] and [3] between genetic algorithms outcomes and model parameters themselves have been here compared. From engineering point of view, it is interesting to investigate also the performance of altogether twelve GA, namely six kinds of SGA and six kinds of MpGA, after ICrA implementation. That is why, in addition, the performance of altogether twelve SGA-MpGA have been presented and included in the comparison.

2 Problem Formulation

Model parameter identification of a considered here *S. cerevisiae* fed-batch fermentation process is performed using real data from on-line and off-line measurements, carried out in the Institute of Technical Chemistry, Hanover, Germany. The details about the process conditions and experimental data set could be found in [18].

According to the mass balance and considering mixed oxidative functional state [18], non-linear mathematical model of *S. cerevisiae* fed-batch fermentation process is commonly described as follows:

$$\frac{dX}{dt} = \left(\mu_{2S} \frac{S}{S + k_S} + \mu_{2E} \frac{E}{E + k_E} \right) X - \frac{F_{in}}{V} X \right) \tag{1}$$

$$\frac{dS}{dt} = \left(-\frac{\mu_{2S}}{Y_{SX}} \frac{S}{S + k_S} X + \frac{F_{in}}{V} (S_{in} - S) \right) \tag{2}$$

$$\frac{dE}{dt} = \left(-\frac{\mu_{2E}}{Y_{EX}} \frac{E}{E + k_E} X + \frac{F_{in}}{V} E \right) \tag{3}$$

$$\frac{dV}{dt} = F_{in} \tag{4}$$

where X is the biomass concentration, [g/l]; S – substrate concentration, [g/l]; E – ethanol concentration, [g/l]; F_{in} – feeding rate, [l/h]; V – bioreactor volume, [l]; S_{in} – substrate concentration in the feeding solution, [g/l]; μ_{2S}, μ_{2E} – maximum values of the specific growth rates, [1/h]; k_S, k_E – saturation constants, [g/l]; Y_{SX}, Y_{EX} – yield coefficients, [-]. All functions are continuous and differentiable and all model parameters fulfil the requirement for non-zero division.

For the considered here model (Eqs. 1–4), the following vector including six model parameters is going to be identified: $p = [\mu_{2S}, \mu_{2E}, k_S, k_E, Y_{SX}, Y_{EX}]$.

As an optimization criterion, mean square deviation between the model output and the experimental data for biomass, substrate and ethanol obtained during the cultivation, has been used:

$$J = \sum_{i=1}^{m} \left(X_{\exp}(i) - X_{\mathrm{mod}}(i) \right)^2 + \sum_{i=1}^{n} \left(S_{\exp}(i) - S_{\mathrm{mod}}(i) \right)^2$$
$$+ \sum_{i=1}^{l} \left(E_{\exp}(i) - E_{\mathrm{mod}}(i) \right)^2 \to \min \tag{5}$$

where m, n and l are the experimental data dimensions; $X_{\exp}$, $S_{\exp}$ and $E_{\exp}$ are the available experimental data for biomass, substrate and ethanol; X_{mod}, S_{mod} and E_{mod} are the model predictions for biomass, substrate and ethanol with a given model parameter vector.

3 Simple and Multi-population Genetic Algorithms for Parameter Identification of *S. cerevisiae* Fed-Batch Cultivation

Genetic algorithms, firstly proposed by Holland [15] and later upgraded by Goldberg [14], are one of the metaheuristic methods based on biological evolution. Frequently used as an alternative to the conventional optimization techniques, both simple and multi-population GA have been successfully applied for different problems solving [13, 14, 18, 19]. Standard simple and multi-population GA use three main genetic operators in a sequence selection, crossover and mutation. This algorithms are here denoted as SGA-SCM and MpGA-SCM, coming from the operators execution order selection, crossover, mutation. SGA-SCM starts with a creation of a randomly generated initial population. On the next stage each solution is evaluated and assigned a fitness value, according which the most suitable solutions are selected. Then, crossover proceeds to form a new offspring. After that mutation with determinate probability is applied aiming to prevent falling of all solutions into a local optimum. While SGA searches a global optimal solution using one population of coded parameter sets, MpGA as more similar to nature working with a number of populations, called subpopulations. After certain number of generations, called isolation time, individuals migrate between the subpopulations. GA terminates when some criterion, such as number of generations reached, or evolution time passed, or fitness threshold reached, or fitness convergence satisfied, etc., is fulfilled. For the purposes of this investigation, GA terminates when a certain number of generations is reached.

While the main idea of GA is to imitate the processes occurring in nature, one can assume that the probability crossover to come first and then mutation is comparable to that both processes to occur in a reverse order; or selection to be performed before or after crossover and mutation, no matter of their order. Following this idea, five modifications of SGA-SCM and five modifications of MpGA-SCM, with different sequence of execution of main genetic operators, have been developed aiming to improve model accuracy and algorithms convergence time [1, 2, 4, 5, 17]. The modifications, namely SGA-SMC and MpGA-SMC (selection, mutation, crossover), SGA-CMS and MpGA-CMS (crossover, mutation, selection), SGA-MCS and MpGA-MCS (mutation, crossover, selection), SGA-CSM and MpGA-CSM (crossover, selection, mutation) and SGA-MSC and MpGA-MSC (mutation, selection, crossover) have been proposed and basically investigated for parameter identification of a fed-batch cultivation of *S. cerevisiae* in [1, 2, 4, 5, 17]. The performance of mentioned above altogether six SGA and six MpGA, applied to the parameter identification of a *S. cerevisiae* fed-batch cultivation process, is going to be assessed by promising ICrA approach. The results will be compared to ICrA assessment of altogether twelve SGA-MpGA algorithms.

4   Intercriteria Analysis

The theoretical framework of the InterCriteria Analysis approach, based on the apparatuses of index matrices (IM) [9–11] and intuitionistic fuzzy sets (IFS) [6–8], is given in details in [12]. Here, ICrA is briefly presented for a completeness.

The initial IM A, presented in the form of Eq. (6) consists of the criteria C_p (for rows), objects O_q (for columns) and real number elements a_{C_p,O_q} for every p, q $(1 \leq p \leq m, 1 \leq q \leq n)$.

$$A = \begin{array}{c|ccccccc}
 & O_1 & \cdots & O_k & \cdots & O_l & \cdots & O_n \\
\hline
C_1 & a_{C_1,O_1} & \cdots & a_{C_1,O_k} & \cdots & a_{C_1,O_l} & \cdots & a_{C_1,O_n} \\
\vdots & \vdots & \ddots & \vdots & \ddots & \vdots & \ddots & \vdots \\
C_i & a_{C_i,O_1} & \cdots & a_{C_i,O_k} & \cdots & a_{C_i,O_l} & \cdots & a_{C_i,O_n} \\
\vdots & \vdots & \ddots & \vdots & \ddots & \vdots & \ddots & \vdots \\
C_j & a_{C_j,O_1} & \cdots & a_{C_j,O_k} & \cdots & a_{C_j,O_l} & \cdots & a_{C_j,O_n} \\
\vdots & \vdots & \ddots & \vdots & \ddots & \vdots & \ddots & \vdots \\
C_m & a_{C_m,O_1} & \cdots & a_{C_m,O_k} & \cdots & a_{C_m,O_l} & \cdots & a_{C_m,O_n}
\end{array}, \qquad (6)$$

Further, an IM with index sets consisting of the criteria (for rows and for columns) with IF pair elements determining the degrees of correspondence between the respective criteria is constructed. A real number a_{C_p,O_q} is comparable about relation R with the other a-object, so that $R(a_{C_k,O_i}, a_{C_k,O_j})$ is defined for each i, j, k. Let $\overline{R}$ be the dual relation of R in the sense that if R is satisfied, then $\overline{R}$ is not satisfied, and vice versa. For example, if "R" is the relation "$<$", then $\overline{R}$ is the relation "$>$", and vice versa. If $S_{k,l}^{\mu}$ is the number of cases in which $R(a_{C_k,O_i}, a_{C_k,O_j})$ and $R(a_{C_l,O_i}, a_{C_l,O_j})$ are simultaneously satisfied, while $S_{k,l}^{\nu}$ is the number of cases is which $R(a_{C_k,O_i}, a_{C_k,O_j})$ and $\overline{R}(a_{C_l,O_i}, a_{C_l,O_j})$ are simultaneously satisfied, it is obvious, that

$$S_{k,l}^{\mu} + S_{k,l}^{\nu} \le \frac{n(n-1)}{2}.$$

Further, for every k, l, satisfying $1 \le k < l \le m$, and for $n \ge 2$,

$$\mu_{C_k,C_l} = 2\frac{S_{k,l}^{\mu}}{n(n-1)}, \quad \nu_{C_k,C_l} = 2\frac{S_{k,l}^{\nu}}{n(n-1)} \qquad (7)$$

are defined. Therefore, $\langle \mu_{C_k,C_l}, \nu_{C_k,C_l} \rangle$ is an intuitionistic fuzzy pair. Next, the following IM is constructed:

$$\begin{array}{c|ccc}
 & C_1 & \cdots & C_m \\
\hline
C_1 & \langle \mu_{C_1,C_1}, \nu_{C_1,C_1} \rangle & \cdots & \langle \mu_{C_1,C_m}, \nu_{C_1,C_m} \rangle \\
\vdots & \vdots & \ddots & \vdots \\
C_m & \langle \mu_{C_m,C_1}, \nu_{C_m,C_1} \rangle & \cdots & \langle \mu_{C_m,C_m}, \nu_{C_m,C_m} \rangle
\end{array}, \qquad (8)$$

that determines the degrees of correspondence between criteria $C_1, \ldots, C_m$.

The sum $\mu_{C_k,C_l} + \nu_{C_k,C_l}$ is not always equal to 1. The difference

$$\pi_{C_k,C_l} = 1 - \mu_{C_k,C_l} - \nu_{C_k,C_l} \qquad (9)$$

is considered as a degree of uncertainty.

The final step of ICrA is to classify the degrees of correspondence between criteria. Let α, $\beta \in [0; 1]$, $\alpha > \beta$, are the threshold values for comparison of μ_{C_k,C_l} and ν_{C_k,C_l}. In general, the criteria C_k and C_l are respectively:

– in a positive consonance, if $\mu_{C_k,C_l} > \alpha$ and $\nu_{C_k,C_l} < \beta$;
– in a negative consonance, if $\mu_{C_k,C_l} < \beta$ and $\nu_{C_k,C_l} > \alpha$;
– in a dissonance, otherwise.

5 Numerical Results and Discussion

All identification procedures as well as ICrA approach implementation are performed on PC Intel Pentium 4 (2.4 GHz) platform running Windows XP.

Altogether twelve modifications of SGA and MpGA with different execution order of main genetic operators selection, crossover and mutation have been consequently applied to estimate the model parameters (vector p) of the considered model (Eqs. 1–4). Due to the stochastic nature of GA, 30 independent runs for each of the applied here GA have been performed. GA operators and parameters are tuned according to [1].

In terms of ICrA, altogether eight criteria are taken into consideration: objective function value J is considered as C_1; convergence time T – as C_2; specific growth rates μ_{2S} and μ_{2E} – respectively as C_3 and C_4; saturation constants k_S and k_E – respectively as C_5 and C_6; yield coefficients Y_{SX} and Y_{EX} – respectively as C_7 and C_8. In [16] and [3] six modifications of SGA and MpGA are considered as objects, respectively. Here, when SGA-MpGA performance have been investigated, altogether twelve objects are investigated. Object O_1 corresponds to SGA-SCM, O_2 – to SGA-SMC; O_3 – to SGA-CMS; O_4 – to SGA-CSM; O_5 – to SGA-MSC; O_6 – to SGA-MCS; O_7 – to MpGA-SCM, O_8 – to MpGA-SMC; O_9 – to MpGA-CMS; O_{10} – to MpGA-CSM; O_{11} – to MpGA-MSC and O_{12} – to MpGA-MCS. For convenience, forenames of objective function, convergence time, fermentation process model parameters and SGA and MpGA modifications are further used instead of a criterion C_i or an object O_j. IMs (10–15) present as follows: the average estimates (10–11), the best ones (12–13), and the worst ones (14–15), respectively for SGA and MpGA, of the values of objective function J, the algorithm convergence time T, [s], as well as of all model parameters towards vector p. Six IMs list the objective function values rounded to the fourth digit after the decimal point, while the rest criteria to the second digit after the decimal point. However, at the step of ICrA implementation, all parameter estimates are used as they have been obtained as a result from parameter identification procedures, in order to be distinguishable and the degrees of uncertainty to be decreased.

$$A_1(SGA, average) =$$

	SGA SCM	SGA SMC	SGA CMS	SGA CSM	SGA MSC	SGA MCS
J	0.02207	0.02210	0.02221	0.02207	0.02222	0.02214
T	215.85	198.06	218.32	232.75	203.55	216.47
μ_{2S}	0.99	0.98	0.92	0.96	0.98	0.99
μ_{2E}	0.15	0.14	0.11	0.15	0.11	0.12
k_S	0.14	0.13	0.11	0.14	0.13	0.13
k_E	0.80	0.80	0.80	0.80	0.80	0.80
Y_{SX}	0.40	0.40	0.42	0.40	0.42	0.41
Y_{EX}	2.02	1.81	1.41	2.02	1.41	1.64

(10)

$$A_2(MpGA, average) =$$

	MpGA SCM	MpGA SMC	MpGA CMS	MpGA CSM	MpGA MSC	MpGA MCS
J	0.02206	0.02195	0.02203	0.02210	0.02203	0.02196
T	113.91	164.40	370.94	317.64	214.06	265.50
μ_{2S}	0.90	0.90	0.90	0.91	0.90	0.91
μ_{2E}	0.14	0.15	0.15	0.12	0.12	0.15
k_S	0.15	0.15	0.15	0.15	0.15	0.15
k_E	0.80	0.84	0.80	0.80	0.84	0.83
Y_{SX}	0.40	0.40	0.40	0.41	0.41	0.40
Y_{EX}	1.86	2.00	2.02	1.65	1.60	1.99

(11)

$$A_3(SGA, best) =$$

	SGA SCM	SGA SMC	SGA CMS	SGA CSM	SGA MSC	SGA MCS
J	0.02206	0.02210	0.02213	0.02214	0.02206	0.02213
T	347.45	278.82	411.44	217.78	279.13	264.30
μ_{2S}	1.00	0.98	0.99	0.91	0.99	1.00
μ_{2E}	0.15	0.14	0.12	0.12	0.15	0.12
k_S	0.14	0.13	0.13	0.11	0.14	0.13
k_E	0.80	0.80	0.80	0.80	0.80	0.80
Y_{SX}	0.40	0.40	0.41	0.41	0.40	0.41
Y_{EX}	2.02	1.82	1.65	1.66	2.03	1.65

(12)

$A_4(MpGA, best) =$

	MpGA SCM	MpGA SMC	MpGA CMS	MpGA CSM	MpGA MSC	MpGA MCS
J	0.02203	0.02193	0.02203	0.02204	0.02199	0.02195
T	131.40	177.19	377.92	343.09	226.79	288.79
μ_{2S}	0.90	0.90	0.90	0.90	0.90	0.90
μ_{2E}	0.15	0.15	0.15	0.14	0.12	0.15
k_S	0.15	0.15	0.15	0.15	0.15	0.15
k_E	0.80	0.85	0.80	0.80	0.85	0.84
Y_{SX}	0.40	0.40	0.40	0.40	0.41	0.40
Y_{EX}	2.03	2.00	2.03	1.95	1.64	2.00

$$(13)$$

$A_5(SGA, worst) =$

	SGA SCM	SGA SMC	SGA CMS	SGA CSM	SGA MSC	SGA MCS
J	0.02207	0.02212	0.02218	0.02229	0.02208	0.02220
T	14.55	14.15	14.74	13.65	13.87	16.15
μ_{2S}	1.00	0.99	0.98	0.96	0.94	0.90
μ_{2E}	0.15	0.13	0.12	0.08	0.15	0.12
k_S	0.14	0.13	0.14	0.11	0.13	0.12
k_E	0.80	0.80	0.80	0.80	0.80	0.80
Y_{SX}	0.40	0.41	0.41	0.43	0.40	0.41
Y_{EX}	2.02	1.79	1.58	1.10	2.06	1.57

$$(14)$$

$A_6(MpGA, worst) =$

	MpGA SCM	MpGA SMC	MpGA CMS	MpGA CSM	MpGA MSC	MpGA MCS
J	0.02211	0.02201	0.02204	0.02218	0.02205	0.02198
T	92.30	142.94	347.33	295.43	195.97	253.55
μ_{2S}	0.92	0.90	0.90	0.92	0.90	0.92
μ_{2E}	0.12	0.15	0.15	0.10	0.11	0.15
k_S	0.15	0.15	0.15	0.15	0.15	0.15
k_E	0.80	0.81	0.80	0.80	0.84	0.84
Y_{SX}	0.41	0.40	0.40	0.42	0.41	0.40
Y_{EX}	1.61	2.00	1.97	1.31	1.52	1.98

$$(15)$$

As seen from (10–15), obtained results show similar values for objective function J after application of six modifications of SGA and MpGA for parameter identification of fermentation process model. There is about 1.6% difference between the best among the best results ($J = 0.02193$ for MpGA-SMC, (13)) and the worst among the worst results ($J = 0.02229$ for SGA-CSM, (14)). On the other hand, the convergence time T increases more than 30 times (411.44 for SGA-CMS, (12) towards 13.65 for SGA-CSM, (14)). Such a small deviation of J proves all considered here SGA modifications as equally reliable and it is of user choice to make a compromise between the model accuracy and convergence time.

ICrA approach has been consequently implemented for each of the constructed IMs $A_1(SGA, average)$, $A_2(MpGA, average)$, $A_3(SGA, best)$, $A_4(MpGA, best)$, $A_5(SGA, worst)$ and $A_6(MpGA, worst)$. After ICrA application, twelve IMs with the calculated degrees of agreement and disagreement between investigated criteria have been obtained. IMs themselves are not shown here, but the results from the ICrA implementation for the cases of average, best and worst evaluations have been summarized in Table 1. Obtained results are ranked by μ_{C_k,C_l} values in the case of average evaluations in SGA.

Table 1 Criteria relations sorted by μ_{C_k,C_l} values in the case of average results

Criteria relation	SGA/MpGA/SGA-MpGA		
	Average results	Best results	Worst results
$J - Y_{SX}$	1.00/0.67/0.74	0.93/0.60/0.70	0.93/0.93/0.82
$\mu_{2E} - k_S$	0.93/0.73/0.70	0.73/0.80/0.73	0.67/0.67/0.56
$\mu_{2E} - k_E$	0.93/0.47/0.64	0.67/0.40/0.62	0.67/0.60/0.52
$\mu_{2E} - Y_{EX}$	0.87/0.93/0.94	0.93/0.87/0.94	1.00/1.00/1.00
$k_S - k_E$	0.87/0.47/0.85	0.40/0.47/0.74	0.33/0.40/0.68
$k_S - Y_{EX}$	0.80/0.80/0.70	0.80/0.80/0.73	0.67/0.67/0.56
$k_E - Y_{EX}$	0.80/0.53/0.64	0.60/0.40/0.59	0.67/0.60/0.52
$\mu_{2S} - k_S$	0.73/0.20/0.21	0.73/0.07/0.18	0.73/0.27/0.29
$\mu_{2S} - \mu_{2E}$	0.67/0.33/0.45	0.60/0.13/0.33	0.67/0.47/0.58
$T - Y_{EX}$	0.60/0.60/0.58	0.53/0.47/0.53	0.40/0.40/0.44
$\mu_{2S} - k_E$	0.60/0.47/0.24	0.27/0.60/0.20	0.60/0.47/0.33
$T - k_S$	0.53/0.53/0.55	0.60/0.33/0.42	0.73/0.47/0.82
$\mu_{2S} - Y_{EX}$	0.53/0.27/0.39	0.53/0.07/0.32	0.67/0.47/0.58
$T - \mu_{2E}$	0.47/0.67/0.55	0.60/0.47/0.53	0.40/0.40/0.44
$T - Y_{SX}$	0.47/0.40/0.41	0.33/0.47/0.41	0.60/0.53/0.50
$J - T$	0.47/0.60/0.52	0.27/0.60/0.53	0.53/0.47/0.32
$T - \mu_{2S}$	0.40/0.53/0.45	0.60/0.47/0.58	0.47/0.53/0.29
$T - k_E$	0.40/0.27/0.45	0.53/0.27/0.39	0.07/0.40/0.62
$\mu_{2S} - Y_{SX}$	0.40/0.60/0.56	0.33/0.73/0.59	0.33/0.47/0.45
$J - \mu_{2S}$	0.40/0.53/0.73	0.27/0.47/0.71	0.27/0.53/0.58
$J - k_E$	0.13/0.13/0.09	0.47/0.07/0.12	0.40/0.27/0.27
$k_E - Y_{SX}$	0.13/0.47/0.35	0.40/0.47/0.39	0.33/0.33/0.42
$k_S - Y_{SX}$	0.13/0.33/0.32	0.20/0.20/0.29	0.33/0.40/0.41
$J - k_S$	0.13/0.40/0.15	0.13/0.33/0.11	0.27/0.47/0.26
$J - Y_{EX}$	0.07/0.33/0.27	0.20/0.47/0.33	0.07/0.13/0.24
$Y_{SX} - Y_{EX}$	0.07/0.13/0.14	0.13/0.20/0.15	0.00/0.07/0.06
$J - \mu_{2E}$	0.07/0.40/0.30	0.13/0.53/0.32	0.07/0.13/0.24
$\mu_{2E} - Y_{SX}$	0.07/0.07/0.11	0.07/0.13/0.14	0.00/0.07/0.06

The results have been graphically presented in Fig. 1, aiming at better interpretation. In Fig. 1, C_1 corresponds to the objective function value J; C_2 – convergence time T; C_3 and C_4 – specific growth rates μ_{2S} and μ_{2E}, respectively; C_5 and C_6 – saturation constants k_S and k_E, respectively; C_7 and C_8 – yield coefficients Y_{SX} and Y_{EX}, respectively.

In [16], where SGA have been assessed separately, no pairs with a degree of uncertainty have been observed, while in [3], where MpGA have been assessed separately, there were such criteria pairs, particularly in the case of the best evaluations. Logically, for SGA-MpGA it has been expected again and further appeared, that in this case there are pairs with a degree of uncertainty, again in the case of the best evaluations. The explanation of this fact is that even using a "row data" from parameter identification procedures, there are some equal evaluations for some of model parameters when different MpGA have been applied.

Table 2 presents the scale of consonance and dissonance [20], on which basis each pair of criteria is going to be assessed.

Based on the presented scale, in the current study the GA performance assessments is going to be done in broader ranges, namely: negative consonance – with $\mu \in (0.05; 0.25]$, uniting first 3 rows of Table 2; dissonance – with $\mu \in (0.25; 0.75]$, uniting next 5 rows of Table 2, and positive consonance with $\mu \in (0.75; 1]$, uniting last 3 rows of Table 2. Thus, the following pair dependencies might be outlined

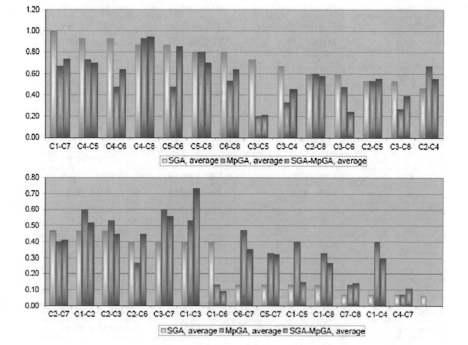

Fig. 1 Degrees of agreement for SGA/MpGA/SGA-MpGA in case of average results

Table 2 Consonance and dissonance scale

Interval of $\mu_{C,C'}$	Meaning
[0–0.05]	Strong negative consonance
(0.05–0.15]	Negative consonance
(0.15–0.25]	Weak negative consonance
(0.25–0.33]	Weak dissonance
(0.33–0.43]	Dissonance
(0.43–0.57]	Strong dissonance
(0.57–0.67]	Dissonance
(0.67–0.75]	Weak dissonance
(0.75–0.85]	Weak positive consonance
(0.85–0.95]	Positive consonance
(0.95–1.00]	Strong positive consonance

for the case of average results of the examined criteria. For the three considered here cases – of an independent assessment of six kinds of SGA, of an independent assessment of six kinds of MpGA, and of assessment of altogether 12 GA (SGA-MpGA), a positive consonance has been observed only for the criteria pair $\mu_{2E} - Y_{EX}$. There are two criteria pairs in a negative consonance $\mu_{2E} - Y_{SX}$, and $Y_{SX} - Y_{EX}$, respectively. The other coincidence has been outlined for $T - Y_{EX}$, $T - \mu_{2E}$, $T - Y_{SX}$, $J - T$, $T - \mu_{2S}$, $\mu_{2S} - Y_{SX}$ and $J - \mu_{2S}$. These altogether seven criteria pairs are in dissonance, no matter weak or strong. When the results of SGA and MpGA performance have been compared, the criteria pair $\mu_{2E} - Y_{EX}$ is the only one observed with the maximum value (i.e. $\mu = 1$) for the degree of agreement, even though it is in the case of worst results. The same criteria pair hits the upper boundary of the μ values in the case of worst results, as such showing a strong positive consonance. A positive consonance have been observed for two criteria pairs, $\mu_{2E} - Y_{EX}$ and $J - Y_{SX}$, respectively for average and best results and only for the worst results.

Another coincidences in SGA and MpGA performance have been observed for the pair $J - k_E$. A negative consonance could be seen for $J - k_E$ in the case of average results. A negative consonance (from strong to weak) for $Y_{SX} - Y_{EX}$ and $\mu_{2E} - Y_{SX}$ could be seen in the cases of average, best and worst results, respectively.

As it could also be seen from Table 1, there are 55 (more than 65% of the evaluations, 84 altogether) ordered triples for the three considered here cases of separately or jointly assessed SGA or/and MpGA, nevermind the cases of average, best or worst results, that assess the corresponding pair indesputably to one of the defined broader ranges. As such, e.g. the pair $\mu_{2E} - Y_{EX}$ is evaluated for all algorithms in all of the cases as in positive consonance, that makes this fact undeniable. There are eight more such coincedences, e.g. $T - Y_{EX}$, $T - \mu_{2E}$, $J - T$, $T - \mu_{2S}$, $\mu_{2S} - Y_{SX}$, $J - \mu_{2S}$ are indespitable assessed as in dissonance, while pairs $Y_{SX} - Y_{EX}$ and $\mu_{2E} - Y_{SX}$ – as in negative consonance. Among those mentioned evaluations, there are a few with a very very close evaluations, even equal in the case ot $\mu_{2E} - Y_{EX}$ in worst results, e.g. $T - Y_{EX}$ in average, best and worst results, $T - k_S$ in average results, $T - \mu_{2E}$ in worst results, etc. This fact is going to demonstrate both the sustainability of the applied GA, SGA or/and MpGA, as well as of the ICrA evaluations.

Distribution of dependencies between criteria pairs according to Table 2, for the cases of average, best and worst evaluations are listed in Table 3.

As shown in Table 3, the only one obvious coincidence is in the case of worst results of SGA, MpGA and SGA-MpGA, where, in any of cosidered cases, there are four criteria pairs in negative consonance, but not always one and the same. As it was mentioned above, only the pairs $Y_{SX} - Y_{EX}$ and $\mu_{2E} - Y_{SX}$ are in negative consonance for all of the applied algorithms in all of the considered cases – average, best and worst. Almost equal results have been obtained for the worst results, that means that the wosrt results are worst no matter what kind of GA is applied. The main reason for luck of other coincidences might be found in the stochastic nature of GA. If one look deeper in the results presented in Table 3, results obtained from the simultaneous assessment of all considered here twelve GA are closer to those, obtained when MpGA had been assessed independantly, then to those obtained when SGA had been assessed independantly. Curiosly, the bigger discrepancies are observed in the case of average evaluations. Thus one can find an another prove of the statement that MpGA are a bit "stronger" but keeping in mind bigger convergence time needed a good solution to be found.

Presented here analysis shows undesputable help of ICrA in the comparison of the performance of different kinds of GA, among them SGA and MpGA, separately and jointly. There is no reason to put away the results obtained when SGA had been assessed independantly, due to the fact that the bigger number of pairs in positive consonance, recorded in this case, gives more knowledge about the considered object, namely the relations between the model parameters of the fermentation process model. Thus having in mind both obtained ICrA estimations in the cases of average, best and worst results, as well as the stochastic nature of GA, it is more reasonable to rely with higher credibility on the results in the case of average values than to results obtained in another two cases. The choice of SGA or MpGA is as always at the hand of user, making a compromise about the convergence time for almost equal model accuracy.

Table 3 Distribution of dependences between criteria pairs

Meaning	SGA/MpGA/SGA-MpGA		
	Average results	Best results	Worst results
Positive consonance	7/2/1	3/3/1	2/2/3
Dissonance	13/22/21	19/18/21	22/22/21
Negative consonance	8/4/6	6/7/6	4/4/4

6 Conclusion

Promising ICrA approach has been here implemented to examine the performance of altogether twelve modifications of GA - six of them of simple and six of them of multi-population genetic algorithms, applied for the purposes of parameter identification procedure of a fermentation process model. After applying SGA and MpGA to a parameter identification of *S. cerevisiae* fed-batch cultivation process, three case studies have been examined- of average, best and worst results with regard to chosen criteria. ICrA approach assisted in establishing of existing relations between fermentation process model parameters and SGA or MpGA outcomes, such as objective function value and convergence time. For the completeness, the performance of altogether twelve SGA-MpGA algorithms have been also calculated, applying ICrA. The results have been compared and thoroughly analyzed. Obtained additional knowledge for relations between model parameters and algorithms outcomes might be useful for improving the model accuracy and the performance of optimization algorithms in further parameter identification procedures.

Acknowledgements The work is partially supported by the National Science Fund of Bulgaria under Grant DM-07/1 "Development of New Modified and Hybrid Metaheuristic Algorithms" and Grant DN-02/10 "New Instruments for Knowledge Discovery from Data, and Their Modelling".

References

1. Angelova, M., Pencheva, T.: Genetic operators significance assessment in multi-population genetic algorithms. Int. J. of Metaheuristics **3**(2), 162–173 (2014)
2. Angelova, M., Tzonkov, S., Pencheva, T.: Modified multi-population genetic algorithm for yeast fed-batch cultivation parameter identification. Int. J. Bioautomation **13**(4), 163–172 (2009)
3. Angelova, M., Pencheva, T.: InterCriteria analysis of multi-population genetic algorithms performance. Ann. Comp. Sci. Inf. Syst. **13**, 77–82 (2017)
4. Angelova, M., Melo-Pinto, P., Pencheva, T.: Modified simple genetic algorithms improving convergence time for the purposes of fermentation process parameter Identification. WSEAS Trans. Syst. **11**(7), 256–267 (2012)
5. Angelova, M., Tzonkov, S., Pencheva, T.: Genetic algorithms based parameter identification of yeast fed-batch cultivation. Lect. Notes Comput. Sci. **6046**, 224–231 (2011)
6. Atanassov, K.: On Intuitionistic Fuzzy Sets Theory. Springer, Berlin (2012)
7. Atanassov, K.: Intuitionistic Fuzzy Sets, VII ITKR Session, Sofia, 20–23 June 1983. Reprinted. Int. J. Bioautomation **20**(S1), S1–S6 (2016)
8. Atanassov, K.: Review and new results on intuitionistic fuzzy sets. Mathematical foundations of artificial intelligence seminar, Sofia, 1988, Preprint IM-MFAIS-1-88. Reprinted: Int. J. Bioautomation **20**(S1), s7–s16 (2016)
9. Atanassov, K.: Generalized Index Matrices. Compt. rend. Acad. Bulg. Sci. **40**(11), 15–18 (1987)
10. Atanassov, K.: On index matrices, part 1: standard cases. Adv. Studies Contemporary Mathe **20**(2), 291–302 (2010)
11. Atanassov, K.: On index matrices, part 2: intuitionistic fuzzy case. Proc. Jangjeon Mathe. Soc. **13**(2), 121–126 (2010)

12. Atanassov, K., Mavrov, D., Atanassova, V.: Intercriteria decision making: a new approach for multicriteria decision making, based on index matrices and intuitionistic fuzzy sets. Issues Intuitionistic Fuzzy Sets and Generalized Nets **11**, 1–8 (2014)
13. Ghaheri, A., Shoar, S., Naderan, M., Hoseini, S.S.: The applications of genetic algorithms in medicine. Oman. Med. J. **30**(6), 406–416 (2015)
14. Goldberg, D.E.: Genetic Algorithms in Search. Optimization and Machine Learning, Addison Wesley Longman, London (2006)
15. Holland, J.: Adaptation in natural and artificial systems: an introductory analysis with application to biology. University of Michigan Press, Control and Artificial Intelligence (1975)
16. Pencheva, T., Angelova, M.: Intercriteria analysis of simple genetic algorithms performance. Advanced Computing in Industrial Mathematics, vol. 681 of studies in computational intelligence, pp. 147–159 (2017)
17. Pencheva, T., Angelova, M.: Modified multi-population genetic algorithms for parameter identification of yeast fed-batch cultivation. Bulg. Chem. Comm. **48**(4), 713–719 (2016)
18. Pencheva, T., Roeva, O., Hristozov, I.: Functional State Approach to Fermentation Processes Modelling. Prof. Marin Drinov Academic Publishing House, Sofia (2006)
19. Roeva, O. (ed.) Real-world application of genetic algorithms. InTech (2012)
20. Roeva, O., Fidanova, S., Vassilev, P., Gepner, P.: Intercriteria analysis of a model parameters identification using genetic algorithm. Ann. Comp. Sci. Inf. Syst. **5**, 501–506 (2015)
21. Todinova, S., Mavrov, D., Krumova, S., Marinov, P., Atanassova, V., Atanassov, K., Taneva, S.G.: Blood plasma thermograms dataset analysis by means of InterCriteria and correlation analyses for the case of colorectal cancer. Int. J. Bioautomation **20**(1), 115–124 (2016)
22. Zaharieva, B., Doukovska, L., Ribagin, S., Radeva, I.: InterCriteria decision making approach for behterev's disease analysis. Int. J. Bioautomation **22**(2), 2018, in press

Structure Optimization and Learning of Fuzzy Cognitive Map with the Use of Evolutionary Algorithm and Graph Theory Metrics

Katarzyna Poczeta, Łukasz Kubuś and Alexander Yastrebov

Abstract Fuzzy cognitive map (FCM) allows to discover knowledge in the form of concepts significant for the analyzed problem and causal connections between them. The FCM model can be developed by experts or using learning algorithms and available data. The main aspect of building of the FCM model is concepts selection. It is usually based on the expert knowledge. The aim of this paper is to present the developed evolutionary algorithm for structure optimization and learning of fuzzy cognitive map on the basis of available data. The proposed approach allows to select key concepts during learning process based on metrics from the area of graph theory: significance of each node, total value of a node and total influence of the concept and determine the weights of the connections between them. A simulation analysis of the developed algorithm was done with the use of synthetic and real-life data.

1 Introduction

Fuzzy cognitive map (FCM) is a directed weighted graph for representing knowledge [11]. It is an effective tool for modeling dynamic decision support systems [14, 26]. Fuzzy cognitive maps allow to visualize complex systems as a set of key concepts (nodes) and connections (links) between them. The FCM model can be built based on expert knowledge [3, 4]. Experts choose the most significant concepts and determine type and strength of the relationships between them (weights of the connections). Fuzzy cognitive map can be also initialized with the use of learning algorithms [16] and available data. Standard supervised [8] and evolutionary algo-

K. Poczeta (✉) · Ł. Kubuś · A. Yastrebov
Kielce University of Technology, al. Tysiąclecia Państwa Polskiego 7,
25-314 Kielce, Poland
e-mail: k.piotrowska@tu.kielce.pl

Ł. Kubuś
e-mail: l.kubus@tu.kielce.pl

A. Yastrebov
e-mail: a.jastriebow@tu.kielce.pl

© Springer Nature Switzerland AG 2019
S. Fidanova (ed.), *Recent Advances in Computational Optimization*,
Studies in Computational Intelligence 795,
https://doi.org/10.1007/978-3-319-99648-6_8

rithms [14, 24, 25] allow to determine the structure of the FCM model based on all data. For each data attribute new concept is created. Next the weights of the connections are specified during learning process. Fuzzy cognitive maps with the large number of concepts are difficult to analyze and interpret. Moreover, with the growth of the number of concepts, the number of connections between them that should be determined increases quadratically. Several researchers have attempted to develop methods of reduction of fuzzy cognitive map size. In [5] a new approach for reduction of the FCM model complexity by merging related or similar initial concepts into the same cluster of concepts is presented. These clusters can be used then as the real concepts in the reduced FCM model. Concepts clustering technique based on fuzzy tolerance relations was used in [17] for modeling of waste management system. The analysis of the decision making capabilities of the less complex FCM shows that proper concepts reductions make models easier to be used keeping their original dynamic behavior. Also cluster validity indexes were introduced to evaluate Fuzzy Cognitive Map design before training phase in [6]. The resulting FCM models are easy to interpret and properly perform the task of prediction. Homenda et al. [7] introduced a time series modeling framework based on simplified Fuzzy Cognitive Maps using a priori nodes rejection criteria. The obtained results confirmed that this approach for simplifying complex FCM models allows to achieve a reasonable balance between complexity and modeling accuracy. Selvin and Srinivasaraghavan proposed an application of the feature selection techniques to reduce the number of the input concepts of fuzzy cognitive map [21]. The feature selection methods were performed based on the significance of each concept to the output concept. However the influences of the connections between the concepts were not taken into consideration. In [18, 20] the structure optimization genetic algorithm for fuzzy cognitive maps learning was presented. It allows to select the most significant concepts and connections between them based on random generation of possible solutions and the error function that takes into account an additional penalty for highly complexity of FCM during learning process. The usefulness of the developed approach was shown on the example of the one-step ahead time series prediction.

The advantage of the FCM model is its graph-based representation that allows to use various methods and metrics from the area of graph theory to analyze the structure and behavior of the modeled system [27]. In [19], we proposed to use two various metrics to optimize the structure of the FCM model by reducing the number of concepts during learning process. The first metric is the degree of a node. It denotes its significance based on the number of concepts it interacts with (is affected by and it affects) [4]. The second metric is one of the system performance indicators: the total (direct and indirect) influence of the concept [2, 22]. In this paper more detailed analysis of the developed approach based on synthetic and real-life data is presented. We used also an additional metric from the are of graph theory: a total value of the node calculated based on the sum of weights of all incoming and outgoing links [4].

The aim of this paper is to present the evolutionary learning algorithm that allows:

- to reduce the size of the FCM model by selecting the most significant concepts,
- to determine the weights of the connections between concepts,
- to approximate the synthetic and real-life data.

The comparison of the developed approach with the standard one based on the all possible concepts and data error and the previously developed approach based on density and system performance indicators [13] was done. The learning process was performed using two effective techniques for FCMs learning: Real-Coded Genetic Algorithm (RCGA) [24] and Individually Directional Evolutionary Algorithm (IDEA) [12].

The outline of this paper is as follows. Section II briefly describes fuzzy cognitive maps. Section III presents the proposed approach for fuzzy cognitive map learning and concepts selection. In Section IV, the results of the simulation analysis based on synthetic and real-life data are presented. Section V contains the conclusion and further work.

2 Fuzzy Cognitive Maps

Fuzzy cognitive map is a directed weighted graph for representing causal reasoning [11]:

$$< X, W > \tag{1}$$

where $X = [X_1, \ldots, X_n]^T$ is the set of the concepts, n is the number of concepts determining the size of the FCM model, W is the connection matrix, $w_{j,i}$ is the weight of the influence between the jth concept and the ith concept, taking on the values from the range $[-1, 1]$. $w_{j,i} > 0$ means X_j causally increases X_i, $w_{j,i} < 0$ means X_j causally decreases X_i.

Fuzzy cognitive map allows to model behavior of dynamic decision support systems and can be used in a what-if analysis [1]. The values of the concepts determine the state of the FCM model and can be calculated according to the selected dynamic model. In the paper one of the most popular dynamic models was used [24]:

$$X_i(t + 1) = F \left(\sum_{j=1, j \neq i}^{n} w_{j,i} \cdot X_j(t) \right) \tag{2}$$

where $X_i(t)$ is the value of the ith concept at the tth iteration, $i = 1, 2, \ldots, n$, t is discreet time, $t = 0, 1, 2, \ldots, T$. Transformation function $F(x)$ normalizes values of the concepts to a proper range. The most often used function is a logistic one, described as follows [24, 25]:

$$F(x) = \frac{1}{1 + e^{-cx}} \tag{3}$$

where c is a parameter, $c > 0$.

3 Proposed Approach

The aim of the proposed approach is structure optimization of fuzzy cognitive map by automatic concepts selection during learning process with the use of metrics from the area of graph theory. This approach requires determination of the decision (output) concepts. Other concepts are input concepts. The obtained model consists only key input concepts that affect to the decision/output concept (or concepts). During learning process we evaluate the candidate FCMs based on data error calculated for decision concepts. The significance of the concept (the degree of the concept), the total value of the node and the total influence of the concept were taken into account in the process of key concepts selection.

The proposed approach contains the following steps:

STEP 1. Initialize random population.

An initial population is generated before starting evolution loop. Each candidate FCM is described by the two vectors. The first vector (4) describes values of weights between concepts [24]:

$$W' = [w_{1,2}, \ldots, w_{1,n}, w_{2,1}, w_{2,3}, \ldots, w_{2,n}, \ldots, w_{n,n-1}]^T \tag{4}$$

where $w_{j,i} \in [-1, 1]$ is the weight of the connection between the jth and the ith concept, $j = 1, 2, \ldots, n$ and n is the number of concepts.

The second vector (5) describes the state of each concept:

$$C = [c_1, c_2, \ldots, c_n,]^T$$
$$c_i \in \{AS, IAS, AAS\} \tag{5}$$

where c_i is the state of ith concept and n is the number of concepts.

Each concept can be in the one of the three states: active (AS), inactive (IAS) and always active (AAS). The decision concept is always active. This mean, that obtained model always contains decision concept (concepts). The concepts with AS state and the decision concepts creates the collection of key concepts.

During the first step, the elements of the W' vector are initialized with the random values form the interval $[-1, 1]$. The state for every node is active for all individual in the initial population. For this reason, the elements of the C vector are equal to AAS for the decision concept (concepts) and AS for the other concepts.

STEP 2. Evaluate population.

Each individual is evaluated based on the following fitness function:

$$fitness(Error) = -Error \tag{6}$$

where *Error* is the objective function calculated on the basis of data error for the decision concepts:

$$Error = \sum_{p=1}^{P_L} \sum_{t=1}^{T_L} \sum_{i=1}^{n_d} |Z_i^p(t) - X_i^p(t)| \tag{7}$$

where $X_i^p(t)$ is the value of the ith decision concept at iteration t of the candidate FCM started from the pth initial state vector, $Z_i^p(t)$ is the value of the ith decision concept at iteration t of the input model started from the pth initial state vector, $t = 0, 1, 2, \ldots, T_L$, T_L is the input learning data length, $i = 1, \ldots, n_d$, n_d is the number of decision concepts, $p = 1, 2, \ldots, P_L$, P_L is the number of the initial learning state vectors.

STEP 3. Check stop condition.

If the number of iterations is greater than *iteration*$_{max}$ then the learning process is stopped.

STEP 4. Select new population.

The temporary population is created from a current base population using roulette-wheel selection with dynamic linear scaling of the fitness function [15].

STEP 5. Select key concepts.

Process of selection of key concepts is carried out in 3 ways:

1. Key concepts are selected at random (SC_RND).
   The state of each input concept for each individual may be changed with a certain probability. The value of state change probability is in the range $(0, 1)$. The concept, whose state is AS may be removed from the key concepts collection by changing the state to IAS. The concept, whose state is IAS may be added to the key concepts collection by changing the state to AS. The values of W' vector are not modified.
2. Key concepts are selected based on the degree of the node (CS_DEG).
   The degree of the node (8) denotes its significance based on the number of concepts it interacts with (is affected by and it affects) [4]:

$$deg_i = \frac{\sum_{j=1,j\neq i}^{n} \theta(w_{i,j}) + \sum_{j=1,j\neq i}^{n} \theta(w_{j,i})}{2n - 1},$$

$$\theta(w_{i,j}) = \begin{cases} 1, & w_{i,j} \neq 0 \\ 0, & w_{i,j} = 0 \end{cases} \tag{8}$$

where n is the number of the concepts; $w_{j,i}$ is the weight of the connection between the jth and the ith concept; $i, j = 1, 2, \ldots, n$.

The structure of the model (states of the concepts) is changed with probability equal to 0.5. If model structure is change, the concept with the lowest value of deg_i from the set of active concepts will be moved to the set of inactive concepts and the concept with the highest value of deg_i from the set of inactive concepts will be moved to the set of inactive concepts.

3. Key concepts are selected based on the total value of the node (CS_VAL).

The total value of the node is calculated based on the sum weights of all incoming links and the sum weights of all outgoing connections (9) [4]:

$$val_i = \frac{\sum_{j=1, j\neq i}^{n} |w_{i,j}| + \sum_{j=1, j\neq i}^{n} |w_{j,i}|}{\sum_{j=1}^{n} \sum_{k=1, k\neq j}^{n} |w_{j,k}|} \tag{9}$$

where n is the number of the concepts; $w_{j,i}$ is the weight of the connection between the jth and the ith concept; $i, j = 1, 2, \ldots, n$.

The structure of the model is changed with probability equal to 0.5. The ith concept with the minimum value of val_i from the set of key concepts will be removed from the key concepts collection. The value of state attribute of this concept will be changed to IAS. The ith concept with maximum value of val_i from concepts whose does not belong to the key concepts collection will be added to the key concepts collection.

4. Key concepts are selected based on total influence of each concept (CS_INF).

The total (direct and indirect) influence between concepts is described as follows [2, 22]:

$$inf_i = \frac{\sum_{j}^{n} (p_{i,j} + p_{j,i})}{2n} \tag{10}$$

where n is the number of the concepts, $p_{j,i}$ is the total (direct and indirect) influence between the jth concept and the ith concept calculated on the basis of the total causal effect path between nodes [13], $i, j = 1, 2, \ldots, n$.

The structure of the model is changed with probability equal to 0.5. If model structure is change, the concept with the lowest value of inf_i from the set of active concepts will be moved to the set of inactive concepts and the concept with the highest value of inf_i from the set of inactive concepts will be moved to the set of inactive concepts.

STEP 6. Apply genetic operators with the use of selected evolutionary algorithm.

In this paper Real Coded Genetic Algorithm [24] and Individually Directed Evolutionary Algorithm were used [12]. The genetic operators were applied only to the W' vector. The C vector was processed by independent procedure described in STEP 5.

STEP 7. Analyze population.

Evolution loop is extended by the process of the analysis of potential solution according to the previously developed approach [13]. The values of weights from $[-0.05, 0.05]$ are rounded down to 0 as suggested in [24]. Next, the matrices with the total influence between concepts $p_{j,i}$ are calculated. If the value of $p_{j,i}$ is in the interval $[-0.1, 0.1]$, the corresponding weight value $w_{j,i}$ is rounded down to 0. Moreover, genetic operators implement density control method of potential solution for consistency of the algorithm. Go to STEP 2.

STEP 8. Choose the best individual and calculate evaluation criteria.

To evaluate performance of the proposed approach, we used two criteria that are commonly used in fuzzy cognitive map learning:

1. Initial error allowing calculation of similarity between the input learning data and the data generated by the FCM model for the same initial state vector:

$$initial_{error} = \frac{1}{P_L \cdot T_L \cdot n_d} \sum_{p=1}^{P_L} \sum_{t=1}^{T_L} \sum_{i=1}^{n_d} |Z_i^p(t) - X_i^p(t)| \qquad (11)$$

where $X_i^p(t)$ is the value of the ith decision concept at iteration t of the candidate FCM started from the pth initial state vector, $Z_i^p(t)$ is the value of the ith decision concept at iteration t of the input model started from the pth initial state vector, $t = 0, 1, 2, \ldots, T_L$, T_L is the input learning data length, $i = 1, \ldots, n_d$, n_d is the number of decision concepts, $p = 1, 2, \ldots, P_L$, P_L is the number of the initial learning state vectors.

2. Behavior error allowing calculation of similarity between the input testing data and the data generated by the FCM model for the same initial state vectors:

$$behavior_{error} = \frac{1}{P_T \cdot T_T \cdot n_d} \sum_{p=1}^{P_T} \sum_{t=1}^{T_T} \sum_{i=1}^{n_d} |Z_i^p(t) - X_i^p(t)| \qquad (12)$$

where $X_i^p(t)$ is the value of the ith decision concept at iteration t of the candidate FCM started from the pth initial state vector, $Z_i^p(t)$ is the value of the ith decision concept at iteration t of the input model started form the pth initial state vector, $t = 0, 1, 2, \ldots, T_T$, T_T is the input testing data length, $i = 1, \ldots, n_d$, n_d is the number of decision concepts, $p = 1, 2, \ldots, P_T$, P_T is the number of the initial testing state vectors.

4 Experiments

To analyze the performance of the developed evolutionary algorithm for structure optimization and learning of fuzzy cognitive maps, synthetic and real-life data were used. The aim of the analysis is to select the most significant concepts, determine the influence between them and approximate the available data for the output concepts.

Standard approach for fuzzy cognitive maps learning (STD), the approaches: for random concepts selection (CS_RND), for selection based on the degree of the concept (CS_DEG), for selection based on the total value of the node (CS_VAL), for selection based on the total influence of the concept (CS_INF) and two previously analyzed algorithms based on density (DEN) [13] and based on system performance indicators (SPI) [13] were compared.

4.1 Dataset

Synthetic data were obtained based on 3 randomly generated FCM models: with 5 concepts (1 output X_5) and density 40% (S1), with 10 concepts (2 outputs X_9, X_{10}) and density of 40% (S2), with 15 concepts (3 outputs X_{13}, X_{14}, X_{15}) and density of 40% (S3).

Real-life data were obtained based on the three FCMs reported in literature [9, 10, 23]. The first real-life model is an intelligent intrusion detection system (R1) [23]. It contains 6 concepts: high login failure (X_1), machine alert (output X_2), user alert (output X_3), login failure same machine different user (X_4), login failure same machine same user (X_5) and login failure different machine same user (X_6). The second model is a weather forecasting system (R2) [10]. It contains 9 concepts: clear sky (X_1), sunny day (output X_2), summer (X_3), monsoon (X_4), rainy day (output X_5), rain (X_6), cloud (X_7), cold day (output X_8), winter (X_9). The last fuzzy cognitive map for public health issues (R3) [9] consists of 9 concepts: city population (X_1), migration into city (X_2), modernization (X_3), garbage per area (X_4), healthcare facilities (X_5), number of diseases per 1000 residents (output X_6), bacteria per area (output X_7), infrastructure improvement (X_8) and sanitation facilities (X_9).

The input data for the learning process were generated starting from the 10 random initial vectors for every map ($P_L = 10$). The resulting FCM models were tested on the basis of 10 testing state vectors ($P_T = 10$) and evaluated with the use of criteria (11)–(12) and the number of concepts n_c.

4.2 Learning Parameters

The following parameters were used for the RCGA algorithm:

- selection method: roulette wheel selection with linear scaling
- recombination method: uniform crossover,
- crossover probability: 0.75,
- mutation method: non-uniform mutation,
- mutation probability: 0.04,
- population size: 100,
- number of elite individuals: 10,
- maximum number of iterations: 100.

Table 1 Average results with synthetic and real-life data

Model	Approach	IDEA			RCGA		
		n_c	$initial_{error}$	$behavior_{error}$	n_c	$initial_{error}$	$behavior_{error}$
S1	STD	5	0.0060 ± 0.0011	0.0069 ± 0.0026	5	0.0024 ± 0.0005	0.0032 ± 0.0006
	DEN	5	0.0034 ± 0.0013	0.0042 ± 0.0014	5	0.0018 ± 0.0005	0.0024 ± 0.0010
	SPI	5	0.0042 ± 0.0022	0.0050 ± 0.0024	5	0.0024 ± 0.0009	0.0029 ± 0.0012
	CS_RND	4.3	0.0048 ± 0.0014	0.0050 ± 0.0017	4	0.0025 ± 0.0007	0.0033 ± 0.0011
	CS_DEG	**3.5**	0.0032 ± 0.0018	**0.0033 ± 0.0016**	3.7	**0.0015 ± 0.0002**	**0.0017 ± 0.0005**
	CS_VAL	3.8	**0.0031 ± 0.0020**	**0.0033 ± 0.0016**	3.9	0.0019 ± 0.0004	0.0026 ± 0.0009
	CS_INF	**3.5**	0.0035 ± 0.0023	0.0047 ± 0.0021	**3.4**	0.0037 ± 0.0026	0.0044 ± 0.0024
S2	STD	10	0.0141 ± 0.0021	0.0179 ± 0.0034	10	0.0112 ± 0.0017	0.0147 ± 0.0030
	DEN	10	0.0106 ± 0.0020	0.0125 ± 0.0036	10	**0.0079 ± 0.0012**	0.0097 ± 0.0007
	SPI	10	0.0115 ± 0.0026	0.0115 ± 0.0026	10	0.0081 ± 0.0012	0.0093 ± 0.0018
	CS_RND	9.3	0.0132 ± 0.0010	0.0171 ± 0.0030	8.1	0.0109 ± 0.0014	0.0147 ± 0.0034
	CS_DEG	8.7	0.0113 ± 0.0025	0.0121 ± 0.0028	**7.3**	0.0088 ± 0.0010	0.0111 ± 0.0023
	CS_VAL	**8**	**0.0095 ± 0.0016**	**0.0105 ± 0.0019**	7.7	0.0085 ± 0.0011	0.0095 ± 0.0013
	CS_INF	9	0.0117 ± 0.0023	0.0120 ± 0.0026	7.9	0.0085 ± 0.0008	**0.0092 ± 0.0018**
S3	STD	15	0.0307 ± 0.0035	0.0344 ± 0.0031	15	0.0305 ± 0.0042	0.0380 ± 0.0034
	DEN	15	0.0210 ± 0.0030	0.0247 ± 0.0029	15	0.0207 ± 0.0031	0.0231 ± 0.0028
	SPI	15	**0.0191 ± 0.0034**	**0.0214 ± 0.0027**	15	**0.0183 ± 0.0022**	**0.0204 ± 0.0022**
	CS_RND	13.3	0.0271 ± 0.0036	0.0324 ± 0.0032	12.8	0.0258 ± 0.0035	0.0297 ± 0.0028
	CS_DEG	13.4	0.0217 ± 0.0028	0.0231 ± 0.0041	11.9	0.0200 ± 0.0026	0.0229 ± 0.0021
	CS_VAL	**13.2**	0.0208 ± 0.0014	0.0233 ± 0.0028	12.6	0.0201 ± 0.0016	0.0233 ± 0.0021
	CS_INF	13.6	0.0218 ± 0.0051	0.0243 ± 0.0050	**11.7**	0.0185 ± 0.0017	0.0234 ± 0.0036

(continued)

Table 1 (continued)

Model	Approach	IDEA			RCGA		
		n_c	$initial_{error}$	$behavior_{error}$	n_c	$initial_{error}$	$behavior_{error}$
R1	STD	6	0.0063 ± 0.0014	0.0059 ± 0.0017	6	$\mathbf{0.0035 \pm 0.0006}$	$\mathbf{0.0041 \pm 0.0008}$
	DEN	6	$\mathbf{0.0051 \pm 0.0010}$	$\mathbf{0.0048 \pm 0.0011}$	6	0.0041 ± 0.0010	0.0043 ± 0.0011
	SPI	6	0.0064 ± 0.0015	0.0061 ± 0.0017	6	0.0049 ± 0.0011	0.0049 ± 0.0008
	CS_RND	5.6	0.0057 ± 0.0013	0.0057 ± 0.0010	5.8	0.0046 ± 0.0006	0.0050 ± 0.0010
	CS_DEG	$\mathbf{5.3}$	0.0058 ± 0.0011	0.0054 ± 0.0011	$\mathbf{5}$	0.0046 ± 0.0011	0.0051 ± 0.0013
	CS_VAL	$\mathbf{5.3}$	0.0080 ± 0.0071	0.0075 ± 0.0068	5.5	0.0049 ± 0.0011	0.0051 ± 0.0011
	CS_INF	5.4	0.0061 ± 0.0020	0.0057 ± 0.0019	5.3	0.0050 ± 0.0015	0.0052 ± 0.0015
R2	STD	9	0.0121 ± 0.0017	0.0126 ± 0.0021	9	0.0102 ± 0.0017	0.0110 ± 0.0026
	DEN	9	$\mathbf{0.0077 \pm 0.0011}$	0.0083 ± 0.0014	9	0.0079 ± 0.0010	0.0081 ± 0.0019
	SPI	9	0.0092 ± 0.0009	0.0086 ± 0.0012	9	0.0076 ± 0.0018	$\mathbf{0.0073 \pm 0.0013}$
	CS_RND	8.6	0.0116 ± 0.0005	0.0113 ± 0.0015	8.3	0.0091 ± 0.0020	0.0098 ± 0.0022
	CS_DEG	8.1	0.0087 ± 0.0016	0.0081 ± 0.0017	7.7	0.0077 ± 0.0015	0.0078 ± 0.0016
	CS_VAL	$\mathbf{7.9}$	0.0083 ± 0.0014	$\mathbf{0.0078 \pm 0.0010}$	$\mathbf{6.4}$	0.0078 ± 0.0012	0.0078 ± 0.0009
	CS_INF	8.1	0.0088 ± 0.0017	0.0084 ± 0.0020	7.3	$\mathbf{0.0070 \pm 0.0017}$	$\mathbf{0.0073 \pm 0.0022}$
R3	STD	9	0.0177 ± 0.0036	0.0177 ± 0.0036	9	0.0136 ± 0.0026	0.0171 ± 0.0039
	DEN	9	0.0149 ± 0.0027	0.0141 ± 0.0039	9	$\mathbf{0.0105 \pm 0.0014}$	0.0124 ± 0.0018
	SPI	9	0.0167 ± 0.0032	0.0164 ± 0.0034	9	0.0157 ± 0.0022	0.0146 ± 0.0032
	CS_RND	8.5	0.0162 ± 0.0025	0.0172 ± 0.0036	7.4	0.0127 ± 0.0020	0.0150 ± 0.0045
	CS_DEG	$\mathbf{8}$	$\mathbf{0.0133 \pm 0.0016}$	$\mathbf{0.0115 \pm 0.0023}$	$\mathbf{7.2}$	0.0123 ± 0.0028	0.0127 ± 0.0019
	CS_VAL	8.2	0.0144 ± 0.0032	0.0139 ± 0.0022	7.6	0.0118 ± 0.0021	$\mathbf{0.0115 \pm 0.0027}$
	CS_INF	8.3	0.0136 ± 0.0020	0.0122 ± 0.0010	7.4	0.0159 ± 0.0034	0.0157 ± 0.0046

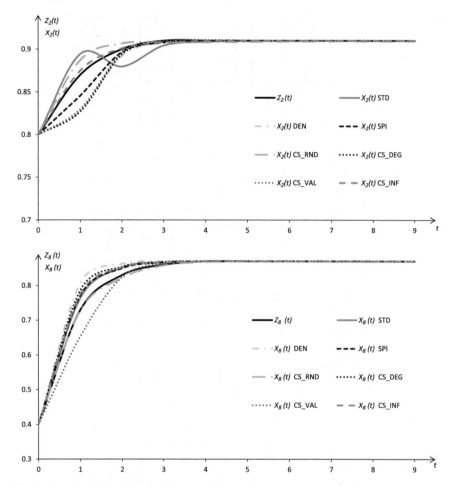

Fig. 1 Sample results of testing of the best FCM models obtained for the analyzed approaches with the use of RCGA algorithm and the real-life data (R2)

The following parameters were used for the IDEA algorithm:

- selection method: roulette wheel selection with linear scaling
- mutation method: directed non-uniform mutation,
- mutation probability: $\frac{1}{n^2-n}$
- population size: 100,
- maximum number of iterations: 100.

Table 2 The number of selections of the concepts

Model	Concept	IDEA				RCGA			
		CS_RND	CS_DEG	CS_VAL	CS_INF	CS_RND	CS_DEG	CS_VAL	CS_INF
S1	X_1	8	8	8	4	7	6	8	8
	X_2	10	10	10	10	10	10	10	7
	X_3	9	4	4	6	7	6	6	5
	X_4	6	3	6	5	6	5	5	4
	X_5	10	10	10	10	10	10	10	10
S2	X_1	8	8	8	9	8	9	8	7
	X_2	10	10	9	7	7	5	8	7
	X_3	10	9	6	8	8	5	8	9
	X_4	9	9	8	10	6	7	7	8
	X_5	10	7	8	9	7	6	5	6
	X_6	9	8	6	9	9	6	7	8
	X_7	9	8	8	9	8	8	5	6
	X_8	8	8	7	9	8	7	9	8
	X_9	10	10	10	10	10	10	10	10
	X_{10}	10	10	10	10	10	10	10	10

(continued)

Table 2 (continued)

Model	Concept	IDEA				RCGA			
		CS_RND	CS_DEG	CS_VAL	CS_INF	CS_RND	CS_DEG	CS_VAL	CS_INF
S3	X_1	8	9	8	10	10	5	8	8
	X_2	9	7	10	10	8	8	9	7
	X_3	9	10	8	6	7	8	8	9
	X_4	8	8	7	8	9	8	8	7
	X_5	8	6	6	8	8	9	9	6
	X_6	9	10	10	10	6	6	6	9
	X_7	10	9	8	8	8	8	8	7
	X_8	9	10	8	10	7	8	8	8
	X_9	7	10	10	10	9	8	9	7
	X_{10}	8	9	9	9	9	6	7	6
	X_{11}	10	9	8	9	9	6	8	6
	X_{12}	8	7	10	8	8	9	8	7
	X_{13}	10	10	10	10	10	10	10	10
	X_{14}	10	10	10	10	10	10	10	10
	X_{15}	10	10	10	10	10	10	10	10
R1	X_1	9	9	8	10	10	9	10	8
	X_2	10	10	10	10	10	10	10	10
	X_3	10	10	10	10	10	10	10	10
	X_4	9	8	7	7	9	7	8	6
	X_5	9	8	10	8	10	6	8	10
	X_6	9	8	8	9	9	8	9	9

(continued)

Table 2 (continued)

Model	Concept	IDEA				RCGA			
		CS_RND	CS_DEG	CS_VAL	CS_INF	CS_RND	CS_DEG	CS_VAL	CS_INF
R2	X_1	9	8	7	8	8	9	7	7
	X_2	10	10	10	10	10	10	10	10
	X_3	9	10	9	9	10	8	6	9
	X_4	9	8	9	9	10	8	8	8
	X_5	10	10	10	10	10	10	10	10
	X_6	9	7	9	9	8	7	5	5
	X_7	10	9	7	9	10	7	3	7
	X_8	10	10	10	10	10	10	10	10
	X_9	10	9	8	7	7	8	5	7
R3	X_1	8	9	8	8	8	9	8	9
	X_2	9	9	6	9	7	6	7	8
	X_3	9	8	10	10	8	8	7	7
	X_4	10	9	9	9	7	8	9	9
	X_5	9	8	9	10	8	7	8	7
	X_6	10	10	10	10	10	10	10	10
	X_7	10	10	10	10	10	10	10	10
	X_8	10	9	10	8	8	6	8	5
	X_9	10	8	10	9	8	8	9	9

4.3 Results

10 experiments were performed for every set of the learning parameters. The average number of selected concepts (n_c) and the average values of initial and behavior error with standard deviations (Avg $\pm$ Std) were calculated. Table 1 summarizes the results of the experiments with synthetic (S1, S2, S3) and real-life (R1, R2, R3) data. The highlighted values in bold show in each experiment the lowest average values obtained for the evaluation criteria: initial error, behavior error and the number of concepts. Figure 1 presents the sample results of testing of the best FCM models obtained for the analyzed approaches with the use of RCGA algorithm and the real-life data (R2).

The obtained results show that the developed algorithm for structure optimization and learning of fuzzy cognitive map allows to approximate synthetic and real-life data with satisfactory accuracy comparable to the standard techniques. We can observe that the proposed approaches for key concepts selection (CS_DEG, CS_VAL, CS_INF), in most of the cases, give the lowest or very close to the lowest values of initial and behavior error. The advantage of the developed algorithm is the ability to optimize the structure of the FCM models (reducing the number of concepts n_c) by selecting the most significant concepts using graph theory metrics: significance of each node, total value of a node and total influence of the concept. Table 2 shows which concepts were most often selected in the analyzed approaches (CS_RND, CS_DEG, CS_VAL, CS_INF).

5 Conclusion

This paper presents the evolutionary algorithm for structure optimization and learning of fuzzy cognitive map on the basis of synthetic and real-life data. Graph theory metrics were used to reduce the size of fuzzy cognitive map by selecting the most significant concepts during learning process. Effectiveness of the proposed approach was analyzed with the use of Real-Coded Genetic Algorithm and Individually Directional Evolutionary Algorithm. The experiments confirmed that the developed approach allows to reduce the number of concepts and determine the weights of the connections between them keeping satisfactory level of data error. We plan to extend our research by applying multi-criteria optimization algorithms.

References

1. Aguilar, J.: A survey about fuzzy cognitive maps papers. Int. J. Comput. Cogn. **3**(2), 27–33 (2005)
2. Borisov, V.V., Kruglov, V.V., Fedulov, A.C.: Fuzzy Models and Networks. Publishing House Telekom, Moscow (2004). (in Russian)

3. Buruzs, A., Hatwágner, M.F., Kóczy, L.T.: Expert-based method of integrated waste management systems for developing fuzzy cognitive map. In: Studies in Fuzziness and Soft Computing (2015)
4. Christoforou, A., Andreou, A.S.: A framework for static and dynamic analysis of multi-layer fuzzy cognitive maps. Neurocomputing **232**, 133–145 (2017)
5. Hatwagner, M.F., Koczy, L.T.: Parameterization and concept optimization of FCM models. In: 2015 IEEE International Conference on Fuzzy Systems (FUZZ-IEEE) (2015)
6. Homenda, W., Jastrzebska, A.: Clustering techniques for Fuzzy Cognitive Map design for time series modeling. Neurocomputing **232**, 3–15 (2017)
7. Homenda, W., Jastrzebska, A., Pedrycz, W.: Nodes selection criteria for fuzzy cognitive maps designed to model time series. In: Filev, D., et al. (eds.) Intelligent Systems'2014. Advances in Intelligent Systems and Computing, vol. 323. Springer, Cham (2015)
8. Jastriebow, A., Poczęta, K.: Analysis of multi-step algorithms for cognitive maps learning. Bull. Pol. Acad. Sci. Tech. Sci. **62**(4), 735–741 (2014)
9. Khan, M.S., Quaddus, M.: Group decision support using fuzzy cognitive maps for causal reasoning. Group Decis. Negot. **13**, 463–480 (2004)
10. Konar, A.: Computational Intelligence: Principles, Techniques and Applications. Computational Intelligence & Complexity. Springer, Berlin (2005)
11. Kosko, B.: Fuzzy cognitive maps. Int. J. Man Mach. Stud. **24**(1), 65–75 (1986)
12. Kubuś, Ł.: Individually directional evolutionary algorithm for solving global optimization problems—comparative study. Int. J. Intell. Syst. Appl. (IJISA) **7**(9), 12–19 (2015)
13. Kubuś, Ł., Poczęta, K., Yastrebov, A.: A new learning approach for fuzzy cognitive maps based on system performance indicators. In: 2016 IEEE International Conference on Fuzzy Systems, Vancouver, Canada, pp. 1–7 (2016)
14. Mateou, N.H., Andreou, A.S.: A framework for developing intelligent decision support systems using evolutionary fuzzy cognitive maps. J. Int. Fuzzy Syst. **19**(2), 150–171 (2008)
15. Michalewicz, Z.: Genetic Algorithms + Data Structures = Evolution Programs. Springer, New York (1996)
16. Papageorgiou, E.I.: Learning algorithms for fuzzy cognitive maps—a review study. IEEE Trans. Syst. Man Cybern. Part C Appl. Rev. **42**(2), 150–163 (2012)
17. Papageorgiou, E.I., Hatwágner, M.F., Buruzs, A., Kóczy, L.T.: A concept reduction approach for fuzzy cognitive map models in decision making and management. Neurocomputing **232**, 16–33 (2017)
18. Papageorgiou, E.I., Poczeta, K.: A two-stage model for time series prediction based on fuzzy cognitive maps and neural networks. Neurocomputing **232**, 113–121 (2017)
19. Poczęta, K., Kubuś, Ł., Yastrebov, A.: Concepts selection in fuzzy cognitive map using evolutionary learning algorithm based on graph theory metrics. In: Ganzha, M., Maciaszek, L., Paprzycki, M. (eds.) Communication Papers of the 2017 Federated Conference on Computer Science and Information Systems. ACSIS, vol. 13, pp. 89–94 (2017)
20. Poczęta, K., Yastrebov, A., Papageorgiou, E.I.: Learning fuzzy cognitive maps using structure optimization genetic algorithm. In: 2015 Federated Conference on Computer Science and Information Systems (FedCSIS), Lodz, Poland, pp. 547–554 (2015)
21. Selvin, N.N., Srinivasaraghavan, A.: Dimensionality reduction of inputs for a Fuzzy Cognitive Map for obesity problem. In: International Conference on Inventive Computation Technologies (ICICT) (2016)
22. Silov, V.B.: Strategic Decision-Making in a Fuzzy Environment. INPRO-RES, Moscow (1995). (in Russian)
23. Siraj, A., Bridges, S.M., Vaughn, R.B.: Fuzzy cognitive maps for decision support in an intelligent intrusion detection system. In: 2001 Joint 9th IFSA World Congress and 20th NAFIPS International Conference (2001)
24. Stach, W., Kurgan, L., Pedrycz, W., Reformat, M.: Genetic learning of fuzzy cognitive maps. Fuzzy Sets Syst. **153**(3), 371–401 (2005)
25. Stach, W., Pedrycz, W., Kurgan, L.A.: Learning of fuzzy cognitive maps using density estimate. IEEE Trans. Syst. Man Cybern. Part B **42**(3), 900–912 (2012)

26. Słoń, G.: Application of models of relational fuzzy cognitive maps for prediction of work of complex systems. Lecture Notes in Artificial Intelligence LNAI, vol. 8467, pp. 307–318. Springer (2014)
27. Wilson, R.J.: An Introduction to Graph Theory. Pearson Education, India (1970)

Fuzziness in the Berth Allocation Problem

Flabio Gutierrez, Edwar Lujan, Rafael Asmat and Edmundo Vergara

Abstract The berth allocation problem (BAP) in a marine terminal container is defined as the feasible berth allocation to the incoming vessels. In this work, we present two models of fuzzy optimization for the continuous and dynamic BAP. The arrival time of vessels are assumed to be imprecise, meaning that the vessel can be late or early up to a threshold allowed. Triangular fuzzy numbers represent the imprecision of the arrivals. The first model is a fuzzy MILP (Mixed Integer Lineal Programming) and allow us to obtain berthing plans with different degrees of precision; the second one is a model of Fully Fuzzy Linear Programming (FFLP) and allow us to obtain a fuzzy berthing plan adaptable to possible incidences in the vessel arrivals. The models proposed has been implemented in CPLEX and evaluated in a benchmark developed to this end. For both models, with a timeout of 60 min, CPLEX find the optimum solution to instances up to 10 vessels; for instances between 10 and 45 vessels it find a non-optimum solution and for bigger instants no solution is founded.

1 Introduction

Port terminals that handle containers are called Maritime Container Terminals (MCT) and have different and more complex operations than passenger and dry or liquid bulk ports. MCT generally serve as a transshipment zone between vessels and land vehicles (trains or trucks).

F. Gutierrez (✉)
Department of Mathematics, National University of Piura, Piura, Peru
e-mail: flabio@unp.edu.pe

E. Lujan
Department of Informatics, National University of Trujillo, Trujillo, Peru
e-mail: edwar_ls@hotmail.com

R. Asmat · E. Vergara
Department of Mathematics, National University of Trujillo, Trujillo, Peru
e-mail: rasmat@unitru.edu.pe

E. Vergara
e-mail: evergara@unitru.edu.pe

© Springer Nature Switzerland AG 2019
S. Fidanova (ed.), *Recent Advances in Computational Optimization*,
Studies in Computational Intelligence 795,
https://doi.org/10.1007/978-3-319-99648-6_9

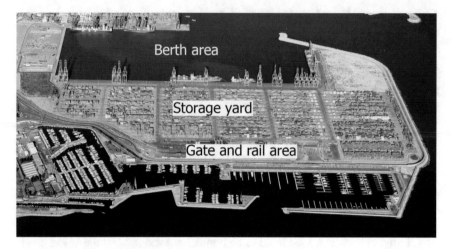

Fig. 1 Container Terminal at Valencia Port

MCT are open systems with three distinguishable areas (see Fig. 1): the berth area, where vessels are berthed for service; the storage yard, where containers are stored as they temporarily wait to be exported or imported; and the terminal receipt and delivery gate area, which connects the container terminal to the hinterland. Each of them presents different planning and scheduling problems to be optimized [14]. For example, berth allocation, quay crane assignment, stowage planning and quay crane scheduling must be managed in the berthing area; the container stacking problem, yard crane scheduling and horizontal transport operations must be carried out in the yard area; and hinterland operations must be solved in landside area.

We will focus our attention on the Berth Allocation Problem (BAP), a well-known NP-Hard combinatorial optimization problem [9], consisting in the allocation for every incoming vessel its berthing position at the quay. Once the vessel arrives to the port, it comes a waiting time to be berthed at the quay. The administrators of MCT must face with two related decisions: where and when the vessels have to be berthed.

The actual times of arrivals for each vessel are highly uncertain depending this uncertainty, e.g., on the weather conditions (rains, storms), technical problems, other terminals that the vessel have to visit and other reasons. The vessels can arrive earlier or later their scheduled arrival time [2, 8]. This situation affects the operations of load and discharge, other activities at the terminal and the services required by costumers. The administrators of MCT change or reviews the plans, but a frequent review of the berthing plan is not a desirable thing from a planning of resources point of view. Therefore, the capacity of adaptation of the berthing plan is important for a good system performance that a MCT manages. As a result, a robust model providing a berthing plan that supports the possible early or lateness in the arrival time of vessels and easily adaptable is desirable. There are many types of uncertainty such as the randomness, imprecision (ambiguity, vagueness), confusion. Many of them can be categorized as stochastic or fuzzy [19]. The fuzzy sets are specially designed to deal with imprecision. There are many attributes to classify the models related to the BAP [1]. The most important are: spatial and temporal. The spatial attribute can be discrete or continuous. In the discrete case, the quay is considered as a finite set of berths, where segments of finite length describe every berth and usually a berth just works for a vessel at once; for the continuous case, the vessels can berth at any position within the limits of the quay. The temporal attribute can be static or dynamical. In the static case, all the vessels are assumed to be at the port before performing the berthing plan; for the dynamical case, the vessels can arrive to the port at different times during the planning horizon. In [1], the authors make an exhaustive review of the literature existing about BAP. To our knowledge, there are very few studies dealing with BAP and with imprecise (fuzzy) data.

A fuzzy MILP (Mixed Integer Lineal Programming) model for the discrete and dynamic BAP was proposed in [4], triangular fuzzy numbers represent the arrival times of vessels, they do not address the continuous BAP. According to Bierwith [1], to design a continuous model, the planning of berthing is more complicated than for a discrete one, but the advantage is a better use of the space available at the quay.

In [5], a MILP fuzzy model for the continuous and dynamic BAP was proposed, this model assigns slacks to support possible delays or earliness of vessels but it also has an inconvenience: if a vessel arrives early or on time, the next vessel has to wait all the time considered for the possible earliness and delay. This represent a big waste of time without the use of the quay and the vessel has to stay longer than is necessary at the port. The evaluation was made just for instances of 8 vessels, which does not allow to evaluate the efficiency of the model.

For this work, the simulation is done in the MCT of the Valencia port, the use of stochastic optimization models is difficult because there are no distributions of probabilities of the delays and advances of the vessels. We assume that the arrival times of vessels are imprecise, for every vessel it is necessary to request the time interval of possible arrival, as well as the more possible time the arrival occurs.

In this work, we present an evaluation of the MILP model to the BAP proposed in [5], with a benchmark to the BAP. We also present a new model fuzzy optimization for the continuous and dynamic BAP, based in Fully Fuzzy Linear Programming.

This paper is organized as follows: In Sect. 2, we describe the basic concepts of fuzzy sets; Sect. 3, presents the formulation of a Fully Fuzzy Linear Programming Problem (FFLP), and describes a method of solution. Section 4, shows the benchmarks for the BAP used to evaluated the models; Sect. 5, shows the evaluation of the fuzzy MIPL model for the BAP. In Sect. 6 we propose, resolve and evaluate a new FFLP model for the BAP. Finally, conclusions and future lines of research are presented in Sect. 7.

2   Introductory Items of Fuzzy Set Theory

The fuzzy sets offers a flexible environment to optimize complex systems. The concepts about fuzzy sets are taken from [17].

2.1 Fuzzy Sets

Definition 1 Let X be the universe of discourse. A fuzzy set $\widetilde{A}$ in X is a set of pairs:

$$\widetilde{A} = \{(x, \mu_{\widetilde{A}}(x)), x \in X\}$$

where $\mu_{\widetilde{A}} : X \rightarrow [0, 1]$ is called the membership function and, $\mu_{\widetilde{A}}(x)$ represents the degree that x belongs to the set $\widetilde{A}$.

In this work, we use the fuzzy sets defined on real numbers R.

Definition 2 The fuzzy set $\widetilde{A}$ in R is normal if $max_x \mu_{\widetilde{A}}(x) = 1$.

Definition 3 The fuzzy set $\widetilde{A}$ in R is convex if and only if the membership function of $\widetilde{A}$ satisfies the inequality

$$\mu_{\widetilde{A}}[\beta x_1 + (1 - \beta)x_2] \geq min[\mu_{\widetilde{A}}(x_1), \mu_{\widetilde{A}}(x_2)] \quad \forall x_1, x_2 \in \text{R}, \quad \beta \in [0, 1]$$

Definition 4 A fuzzy number is a normal and convex fuzzy set in R.

Definition 5 A triangular fuzzy number (see Fig. 2) is represented by $\widetilde{A} = (a_1, a_2, a_3)$.

Definition 6 The fuzzy triangular number $\widetilde{A} = (a1, a2, a3)$ is denominated a non-negative fuzzy triangular number $\Leftrightarrow$ $a_1 \geq 0$.

Definition 7 Let $\widetilde{A}$ a fuzzy set and a real number $\alpha \in [0, 1]$. The crisp set

$$A_\alpha = \{x : \mu_{\widetilde{A}}(x) \geq \alpha, x \in \text{R}\}$$

is called α-*cut* of $\widetilde{A}$ (Fig. 2).

This concept provides a very interesting approach in fuzzy set theory, since the family of α-*cuts* contains all information about the fuzzy set. By adjusting the α

Fig. 2 Interval corresponding to an α-*cut* level, for a triangular fuzzy number

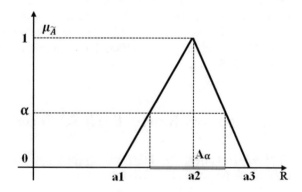

value we can get the range or set of values that satisfy a given degree of membership. In other words, the α value ensures a certain level of satisfaction, precision of the result or robustness of the model.

To a fuzzy set with membership function of triangular type, $\widetilde{A} = (a1, a2, a3)$ (see Fig. 2), the $\alpha - cut$ is given by:

$$A_\alpha = [a1 + \alpha(a2 - a1), a3 - \alpha(a3 - a2)]. \tag{1}$$

2.2 Fuzzy Arithmetic

If we have the nonnegative triangular fuzzy numbers $\widetilde{a} = (a1, a2, a3)$ y $\widetilde{b} = (b1, b2, b3)$, the operations of sum and difference are defined as follows:

Sum: $\widetilde{a} + \widetilde{b} = (a1 + b1, a2 + b2, a3 + b3)$.
Difference: $\widetilde{a} - \widetilde{b} = (a1 - b3, a2 - b2, a3 - b1)$.

2.3 Comparison of Fuzzy Numbers

Comparison of fuzzy numbers allows us to infer between two fuzzy numbers $\widetilde{a}$ and $\widetilde{b}$ to indicate the greatest one. However, fuzzy numbers do not always provide an ordered set as the real numbers do. All methods for ordering fuzzy numbers have advantages and disadvantages. Different properties have been applied to justify comparison of fuzzy numbers, such as: preference, rationality, and robustness [6, 15].

In this work, we use the method called First Index of Yagger [16]. This method uses the ordering function

$$\mathscr{R}(A) = \frac{a1 + a2 + a3}{3} \tag{2}$$

As a result, $A \leq B$ when $\mathscr{R}(A) \leq \mathscr{R}(B)$, that is,

$$a1 + a2 + a3 \leq b1 + b2 + b3.$$

2.4 Distributions of Possibility

The Imprecision can be represented by possibility distributions [18]. These distributions allow us to formalize, in a reliable way, a very large amount of situations estimating magnitudes located in the future. The measure of possibility of an event can be interpreted as the degree of possibility of its occurrence. Among the various types of distributions, triangular and trapezoidal ones are most common. Formally, the distributions of possibility are fuzzy numbers; in this work, we use triangular distributions of possibility $\tilde{a} = (a1, a2, a3)$, which are determined by three quantities: $a2$ is value with the highest possibility of occurrence, $a1$ and $a3$ are the upper and lower limit values allowed, respectively (Fig. 2). These bound values can be interpreted, e.g., as the most pessimistic and the most optimistic values depending on the context.

3   Fully Fuzzy Linear Programming

Fuzzy mathematical programming is useful to handle situations within optimization problems including imprecise parameters [10]. There are different approaches to the fuzzy mathematical programming. When in the problem, the parameters and decision variables are fuzzy and linear, this can be formulated as s FFLP problem. There are many methodologies of solution to a FFLP [3]. Mostly of them, convert the original fuzzy model in a classical satisfactory model.

In this work, we use the method of Nasseri et al. [11]

Given the FFLP problem

$$max \ \sum_{j=1}^{n} \tilde{C}_j \tilde{X}_j$$

Subject to

$$\sum_{j=1}^{n} \tilde{a}_{ij}\tilde{x}_j \leq \tilde{b}_i, \forall i = 1 \ldots m \tag{3}$$

where parameters $\tilde{c}_j, \tilde{a}_{ij}, \tilde{b}_j$ and the decision variable $\tilde{x}_j$ are nonnegative fuzzy numbers $\forall j = 1 \ldots n, \forall i = 1 \ldots m$.

If all parameters and decision variables are represented by triangular fuzzy numbers, $\tilde{C} = (c1_j, c2_j, c3_j)$, $\tilde{a}_{ij} = (a1_{ij}, a2_{ij}, a3_{ij})$, $\tilde{b}_i = (b1_i, b2_i, b3_i)$, $\tilde{x}_j = (x1_j, x2_j, x3_j)$.

Nasseri's Method transforms (3) into a classic problem of mathematical programming.

$$max \ \ \mathscr{R}\left(\sum_{j=1}^{n}(c1_j, c2_j, c3_j)(x1_j, x2_j, x3_j)\right)$$

$$\sum_{j=1}^{n} a1_{ij}x1_j \leq b1_i, \forall i = 1 \ldots m \tag{4}$$

$$\sum_{j=1}^{n} a2_{ij}x2_j \leq b2_i, \forall i = 1 \ldots m \tag{5}$$

$$\sum_{j=1}^{n} a3_{ij}x3_j \leq b3_i, \forall i = 1 \ldots m \tag{6}$$

$$x2_j - x1_j \geq 0, \quad x3_j - x2_j \geq 0 \tag{7}$$

where $\mathscr{R}$ is an ordering function (see Sect. 2.3).

4 Benchmarks for the Berth Allocation Problema

As far as we know, BAPQCAP is the only benchmark presenting instances of problems berthing and crane allocation to the arriving vessels to the TMC. This benchmark is deterministic one since imprecision is nor considered in any of its parameters.

BABQCAP-Imprecise is an extended version of the BAPQCAP, where the arrival times of vessels are considered imprecise, that is, a vessel can be early or delayed. This version has been developed to evaluate the efficiency of the presented models in this paper.

4.1 Benchmark BAPQCAP

The researching group "Inteligencia Artificial - Sistemas de Optimizacion" of the Universidad Politécnica de Valencia (Spain), has developed a benchmark to the BAP and to the Quay Crane Assignment Problem (QCAP). The benchmark is formed by groups of vessels from 5,10,15 to 100; each group consists of 100 instances. This benchmark has been used to evaluate different meta-heuristics to the BAP and the QCAP [12, 13].

Every file consists of 100 vessels describes one for line. To each vessel the following information is available (see Table 1):

id: Identifier of vessel.
l: Length.
a: Arrival time.
h: Number of movement needed to be loaded and unloaded.
pri: Priority of vessels ($1 \leq p \leq 10$), where 10 is the highest priority and 1, the lowest one.

Constraints considered: The length of the quay is 700 m. The number of crane available is 7. The maximum number of cranes allocate to a vessel depend of its

Table 1 Example of one instance to the benchmark BAPQCAP

id	l	a	h	prio
1	260	16	416	9
2	232	31	968	3
3	139	68	364	6
4	193	82	761	6
5	287	105	686	3
6	318	116	630	7
7	366	138	811	2
8	166	157	156	2
9	109	163	783	5
10	251	179	222	1

length. There is a distance of security that must be respected (35 m. between cranes). The maximum number of cranes that can be allocate is 5. The number of movement performed for a crane in a certain time is 2.5.

4.2 Benchmark BAPQCAP-Imprecise

With the aim of simulate the imprecision existing in the arrival time of a vessel to the TMC, in every instance of the benchmark BAPQCAP, the possibility of delay and advance was added to the arrival time up to an allowed tolerance. This possibility is represented by a fuzzy triangular number (a_1, a_2, a_3) (see Fig. 2).

Where:

a1:   Minimum allowed advance in the arrival of the vessel. This value is random and it is generated within the range [a −20, a].

a2:   Arrival time with the higher possibility of a vessel (taken from original benchmark).

a3:   Maximum allowed delay in the arrival of the vessel. This value is random too and it us generated within the range [a, a + 20].

Table 2, shows the modification done to the instance of Table 1. We can appreciate the fourth column is the value of the arrival time of vessel with the highest possibility, the third one represents the advance and the fifth one, the delay.

Table 2 Example of one instance with imprecision in the arrival time of vessels

id	l	a1	a2	a3	h	prio
1	260	9	16	17	416	9
2	232	17	31	33	968	3
3	139	51	68	70	364	6
4	193	80	82	87	761	6
5	287	87	105	113	686	3
6	318	111	116	133	630	7
7	366	127	138	146	811	2
8	166	142	157	157	156	2
9	109	153	163	172	783	5
10	251	168	179	181	222	1

Table 3 Instance with 8 vessels

Barcos	a1	a2	a3	h	l
V1	4	8	34	121	159
V2	0	15	36	231	150
V3	18	32	50	87	95
V4	9	40	46	248	63
V5	32	52	72	213	219
V6	55	68	86	496	274
V7	62	75	90	435	265
V8	45	86	87	146	94

4.3 Case Study

With the aim to show the advantages and disadvantages of the models presents in this work, we use as a case study one instance consisting of 8 vessels (Table 3). In Fig. 3, we show the imprecise arrival of vessel as a triangular fuzzy number.

For example, to the vessel V1, the most possible arrival is at 8 units of time, but it could be early or late up to 4 and 34 units of time, respectively; the handling time is 121 and the length of vessel is 159.

5 A MILP Fuzzy Model for the Continuous and Dynamic BAP

We present the notation used in the model, the model, the solution and the evaluation of the model.

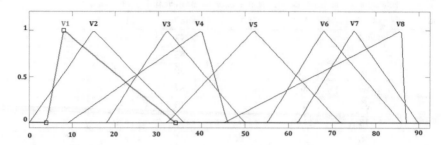

Fig. 3 Imprecise arrival of vessels showed in Table 3

5.1 Notation for the BAP

Figure 4, shows the main parameters used in the models.

   L: Total length of the quay at the MCT.

   H: Planning horizon.

Let V be the set of incoming vessels, the problem data for each vessel $i \in V$ are given by:

   a_i : Arrival time at port.

   l_i : Vessel length.

   h_i : Handling time of the vessel in the berth (service time).

With these data, the decision variables m_i and p_i must be obtained.

   m_i : Berthing time of vessel.

   p_i : Berthing position where the vessel will moor.

With the data and decision variables are obtained w_i and d_i

   $w_i = m_i - a_i$: Waiting time of vessel since the arrival to the berthing.

   $d_i = m_i + h_i$.

Fig. 4 Representation of a vessel according to the time and position

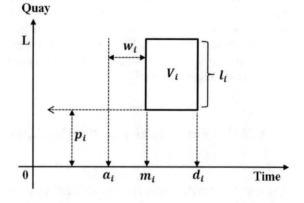

Depending on the model, the arrival times, berthing times, handling time and departure times of the vessel can be considered to be of fuzzy nature (imprecise) and denoted by $\tilde{a}$, $\tilde{m}$, $\tilde{h}$ and $\tilde{d}$, respectively.

We consider the next assumptions: all the information related to the waiting vessels is known in advance, every vessel has a draft that is lower or equal to the draft of the quay, the berthing and departures are not time consuming, simultaneous berthing is allowed, safety distance between vessels is not considered.

5.2 A MILP Fuzzy Model

In this model, we assume imprecision in the arrival time of vessels, meaning that the vessels can be late or early up to an done allowed tolerance.

Formally, we consider that imprecision in the arrival time of vessels is a fuzzy number $\tilde{a}$.

The goal is to allocate a certain time and a place at the quay to every vessel according certain constraints, with the aim of minimize the total waiting time of vessels.

$$T = \sum_{i \in V} (m_i - a_i).$$

Based in the deterministic model proposed in [7], we propose the following model of fuzzy optimization.

$$\min \sum_{i \in V} (m_i - \tilde{a}_i) \tag{8}$$

Subject to:

$$m_i \geq \tilde{a}_i \quad \forall i \in V \tag{9}$$

$$p_i + l_i \leq L \quad \forall i \in V \tag{10}$$

$$p_i + l_i \leq p_j + M(1 - z_{ij}^x) \quad \forall i, j \in V, \ i \neq j \tag{11}$$

$$m_i + h_i \leq H \quad \forall i \in V \tag{12}$$

$$m_j - (m_i + h_i) + M(1 - z_{ij}^y) \geq S(\tilde{a}_i) \quad \forall i, j \in V, \ i \neq j \tag{13}$$

$$z_{ij}^x + z_{ji}^x + z_{ij}^y + z_{ji}^y \geq 1 \quad \forall i, j \in V, \ i \neq j \tag{14}$$

$$z_{ij}^x, z_{ij}^y \in \{0, 1\} \quad \forall i, j \in V, \ i \neq j. \tag{15}$$

If the deterministic and fuzzy parameters are of linear-type we are dealing with a fuzzy MILP model. The constraints are explained below:

- Constraint (9): the berthing time must be at least the same as the fuzzy arrival time.
- Constraint (10): There must be enough space at the quay for the berthing.
- Constraint (11): at the quay, a vessel need to be to left or right side of another one.

- Constraint (12): the berthing plan must be adjusted within the planning horizon.
- Constraint (13): for a vessel j berthing after vessel i, its berthing time m_j must include the time $S(\tilde{a}_i)$ of advance and delay tolerated to the vessel i.
- Constraint (14): the constraints (11) and (13) must be accomplished.

where z_{ij}^x decision variable indicating if the vessel i is located to the left of vessel j at the berthing ($z_{ij}^x = 1$), $z_{ij}^y = 1$ indicates that the berthing time of vessel i is before the berthing time of vessel j. M is a big integer constant.

5.3   Solution of the Model

The imprecise arrival for every vessel is represented by triangular distribution of possibility $\tilde{a} = (a1, a2, a3)$ (see Fig. 2). We consider that arrivals will not occur before $a1$, not after $a3$. The arrival with the maximum possibility is $a2$.

For a triangular fuzzy number $\tilde{a} = (a1, a2, a3)$, according to (1), its α−cut is given by:
$$A_\alpha = [a1 + \alpha(a2 - a1), a3 - \alpha(a3 - a2)]$$

The α−cut represents the time interval allowed for the arrival time of a vessel, given a grade precision α. The size of the interval $S(\alpha) = (1 - \alpha)(a3 - a1)$ must be taken into account to the berthing time vessel next to berth. It can be observed that for the value α, the earliness allowed is $E(\alpha) = (1 - \alpha)(a2 - a1)$, the delay allowed is $D(\alpha) = (1 - \alpha)(a3 - a2)$ and $S(\alpha) = e(\alpha) + D(\alpha)$.

In Fig. 5, the alpha cuts $B1_{0.5}$, $B2_{0.5}$ y $B3_{0.5}$ to the arrival of three vessels, with a level cut $\alpha = 0.5$ are showed.

By using the alpha-cuts as a method of defuzzification to the fuzzy arrival of vessels, a solution to the fuzzy BAP model is obtained with the next auxiliary parametric MILP model.

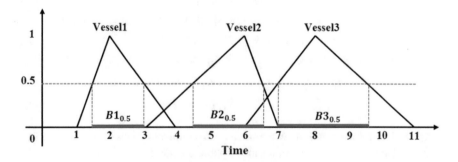

Fig. 5 α-cut for $\alpha = 5$ to the fuzzy arrival of three vessels

Input: Set of incoming vessels V.
Output: Berthing plans to V with different grades of precision
For each $\alpha = \{0, 0.1, \ldots, 1\}$.
   earliness allowed to the vessel i

$$E_i(\alpha) = (1 - \alpha) * (a2_i - a1_i).$$

delay allowed i

$$D_i(\alpha) = (1 - \alpha) * (a3_i - a2_i).$$

tolerance time allowed to the arrival of vessel i

$$S_i(\alpha) = E_i(\alpha) + D_i(\alpha) \quad \forall i \in V$$

$$\min \sum_{i \in V} (m_i - (a1 + \alpha * (a2 - a3))) \tag{16}$$

subject to:

$$m_i \geq (a1 + \alpha * (a2 - a1)) \quad \forall i \in V \tag{17}$$
$$p_i + l_i \leq L \quad \forall i \in V \tag{18}$$
$$p_i + l_i \leq p_j + M(1 - z_{ij}^x) \quad \forall i, j \in V, \ i \neq j \tag{19}$$
$$m_j - (m_i + h_i) + M(1 - z_{ij}^y) \geq S_i(\alpha) \quad \forall i, j \in V, \ i \neq j \tag{20}$$
$$z_{ij}^x + z_{ji}^x + z_{ij}^y + z_{ji}^y \geq 1 \quad \forall i, j \in V, \ i \neq j \tag{21}$$
$$z_{ij}^x, z_{ij}^y \in \{0, 1\} \quad \forall i, j \in V, \ i \neq j. \tag{22}$$

The planning horizon is given by:

$$H = \sum_{i \in V} (h_i) + max\{a3_i, i \in V\}.$$

In the parametric MILP model, the value of α is the grade of precision allowed in the arrival time of vessels. For every $\alpha \in [0, 1]$, and for every vessel i, the tolerance times allowed S_i are computed.

For example, if we have the fuzzy arrival time$(a1, a2, a3) = (22, 32, 47)$.

For $\alpha = 0.8$, the earliness allowed is $E(0.8) = (1 - 0.8)(32 - 22) = 2$; the delay allowed is $D(0.8) = (1 - 0.8)(47 - 32) = 3$ and the total tolerance allowed is $S(0.8) = 5$.

For $\alpha = 0.6$, the earliness allowed is $E(0.6) = (1 - 0.6)(32 - 22) = 4$; the delay allowed is $D(0.6) = (1 - 0.6)(47 - 32) = 6$ and the total tolerance allowed is $S(0.6) = 10$.

The lower the value α is, the lower the precision, that is, the larger the size of the amount of time allowed at the arrival of every vessel.

5.4 Evaluation

To the evaluation we have used a personal computer equipped with a Core (TM) i5 - 4210U CPU 2.4 Ghz with 8.00 Gb RAM. The experiments were performed with a timeout of 60 min.

For each instance, eleven degrees of precision ($\alpha = \{0, 0.1, \ldots, 1\}$), generate eleven berthing planes.

5.4.1 Evaluation of the Case Study

As an illustrative example, to the vessels of Table 3, three different berthing plans are showed in Tables 4, 5 and 6.

For $\alpha = 1$ (maximum precision in the arrival of vessels), in all vessels, the earliness and delays are $E = 0$ and $D = 0$, respectively, that is, earliness and delays are not allowed in the arrival of vessels. In most cases, if a vessel has a delay with respect to its precise arrival time, this plan ceases to be valid. For example, the vessel $V3$ has berthing time $m2 = 32$ and a departure time $d2 = 119$, if this vessel has a delay, the vessel $V8$ can't berth at its allocated time $m2 = 119$, and the next vessel $V7$ also can't berth in its allocated time $m2 = 265$. This can be observed in Fig. 6. To a greater number of vessels (as in real life), the delay of vessels complicate even more the berthing plans.

To precision degree $\alpha = 0.5$, e.g., to the vessel $V3$, the optimum berthing time is $m2 = 32$, the earliness allowed is $E = 7$, the delay allowed $D = 9$, that is, the vessel can berth and the time interval [25, 41], and the departure can be at the time interval [112, 128]. After the vessel $V3$, the vessel $V8$ can berth at the time $m2 = 128$ with

Table 4 Berthing plan to $\alpha = 1$

Barcos	a1	a2	a3	E	D	m1	m2	m3	h	d1	d2	d3	l	p
V1	4	8	34	0	0	8	8	8	121	129	129	129	159	541
V2	0	15	36	0	0	15	15	15	231	246	246	246	150	391
V3	18	32	50	0	0	32	32	32	87	119	119	119	95	233
V4	9	40	46	0	0	40	40	40	248	288	288	288	63	328
V5	32	52	72	0	0	52	52	52	213	265	265	265	219	0
V6	55	68	86	0	0	246	246	246	496	742	742	742	274	426
V7	62	75	90	0	0	265	265	265	435	700	700	700	265	0
V8	45	86	87	0	0	119	119	119	146	265	265	265	94	219

and earliness allowed of $E = 21$ and a delay allowed of $D = 0.5$, the optimum time of berthing is $m2 = 148.5$, but it can berth at the time interval [128, 149] (see Fig. 6).

In $\alpha = 0$ (minimum precision allowed in the arrival time of vessels), the earliness and delays are increased, e.g., to the vessel $V3$, the optimum time of berthing is $m2 = 32$ (the same as for $\alpha = 0.5$), but the earliness allowed is $E = 14$ and the delay allowed is $D = 18$. Therefore, the time interval where the vessel can berth is [18, 50] (see Fig. 6).

Considering the structure of the model created, for every value of α, the earliness and delays allowed are proporcional to its maximum earliness and delay time. For example, for $\alpha = 0.5$, the vessel $V1$ can be early or delayed up to a maximum of 2 and 13 units of time, respectively (see Table 5). If $\alpha = 0.0$, the earliness ad delays to the vessel $V1$, are $E = 4$ y $D = 26$, respectively (see Table 6).

For all values of α, the model was resolver in an optimum way. In Table 7, the objective function T and the computation time used by CPLEX to obtain the the solution to the different degrees of precision α are showed. The lower $T = 401$ is obtained within a time of 2.59 seconds, corresponding to a degree of precision $\alpha = 1$; and the greater $T = 516.90$, is obtained in a time of 2.95 seconds corresponding to a degree of precision $\alpha = 0$.

Table 5 Berthing plan with $\alpha = 0.5$ of precision in the arrival time of vessels

Barcos	a1	a2	a3	E	D	m1	m2	m3	h	d1	d2	d3	l	p
V1	4	8	34	2	13	6	8	21	121	127	129	142	159	219
V2	0	15	36	8	11	7.5	15	26	231	239	246	257	150	392
V3	18	32	50	7	9	25	32	41	87	112	119	128	95	605
V4	9	40	46	16	3	25	40	43	248	273	288	291	63	542
V5	32	52	72	10	10	42	52	62	213	255	265	275	219	0
V6	55	68	86	7	9	256	263	272	496	753	759	768	274	265
V7	62	75	90	7	8	275	282	289	435	710	717	724	265	0
V8	45	86	87	21	1	128	149	149	146	274	295	295	94	606

Table 6 Berthing plan to $\alpha = 0$

Barcos	a1	a2	a3	E	D	m1	m2	m3	h	d1	d2	d3	l	p
V1	4	8	34	4	26	4	8	34	121	125	129	155	159	282
V2	0	15	36	15	21	0	15	36	231	231	246	267	150	441
V3	18	32	50	14	18	18	32	50	87	105	119	137	95	605
V4	9	40	46	31	6	9	40	46	248	257	288	294	63	0
V5	32	52	72	20	20	32	52	72	213	245	265	285	219	63
V6	55	68	86	13	18	267	280	298	496	763	776	794	274	328
V7	62	75	90	13	15	285	298	313	435	720	733	748	265	63
V8	45	86	87	41	1	137	178	179	146	283	324	325	94	606

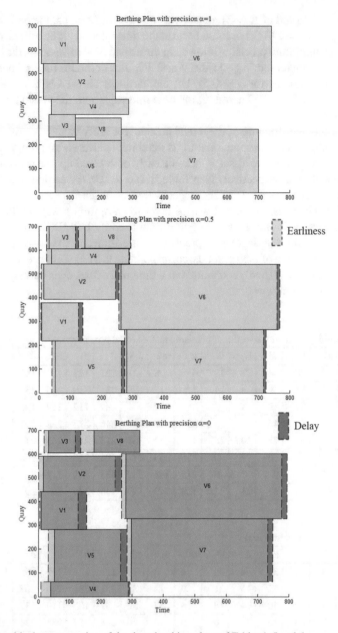

Fig. 6 Graphical representation of the three berthing plans of Tables 4, 5 and 6

There is a linear relationship between α and T, the decrease of α, increases the value of T, e.g., to a degree of precision $\alpha = 0.5$, the value of T is 459.27; and to $\alpha = 0$ is 516.90.

Table 7 Value of the objective function to every degree of precision

α	T	Time (s)
1.00	401.00	2.59
0.90	412.66	2.69
0.80	424.31	2.77
0.70	435.97	2.77
0.60	447.62	2.61
0.50	459.27	2.83
0.40	470.86	2.85
0.30	482.43	2.83
0.20	493.99	2.81
0.10	505.47	2.76
0.00	516.90	2.95

The maker decisions of the TMC, can choose a plan according to the pair (α, T) that is a satisfactory solution.

For example, if a plan using the lower waiting time of vessels is desirable, though earliness and delays in the vessel arrival are not permitted, he can choose the pair (1, 401); if a plan with 0.5 of precision in the arrival of vessels is desirable, though the value of waiting time increases, he has the possibility to choose the pair (0.5, 459.27).

This model assigns slacks to support possible delays or earliness of vessels but it also has an inconvenience: if a vessel arrives early or on time, the next vessel has to wait all the time considered for the possible earliness and delay. This represent a big waste of time without the use of the quay and the vessel has to stay longer than is necessary at the port.

5.4.2 Evaluation of the Benchmark BAPQCAP-Imprecise

Table 8, shows the average of results obtained by CPLEX to the Benchmark BAPQCAP-Imprecise (see Sect. 4.2) with a precision of $\alpha = 0.5$.

The values showed are the average of the objective function of solutions founded ($Avg\ T$), the number of instances solved with optimality (#Opt) and the number of instances solved without certify optimality (#NOpt).

From results, it can be observed that in all the solved cases, T increases as the number of vessels increases. To the given timeout, CPLEX, has found the optimum solution in 59% of the instances with 10 vessels; a non-optimum solution in 100% of the instances from 15 to 45 vessels; and for number of vessels greater or equal to 50 no solution was founded.

The growth of T to the values of $\alpha = \{0, 0.5, 1\}$ is shown in Fig. 7. With the given timeout, CPLEX has found a solution up to instances of 40 vessels to $\alpha = 1$; to $\alpha = 0$

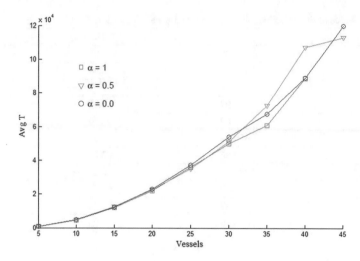

Fig. 7 Evaluation of the imprecise benchmark to different values of α

and $\alpha = 0.5$ up to instances of 45 vessels. For other values of α, CPLEX has found a solution up to instances of 45 vessels.

Table 8 Evaluation of the benchmark BAPQCAP-Imprecise to $\alpha = 0.5$

Vessels	Avs T	# Opt	# NOpt
5	865.7	100	100
10	4595.94	41	59
15	12520.2	0	100
20	22519.99	0	100
25	35039.99	0	100
30	51074.99	0	100
35	72695.99	0	100
40	107216.99	0	100
45	112980.49	0	100
50	–	0	0
55	–	0	0
60	–	0	0

6 A FFLP Model for the Continuous and Dynamic BAP

We present and new model for the continuous and dynamic BAP, that solves the inconvenience of a great waste of time without the use of the quay of the MILP fuzzy model (see Sect. 5).

6.1 A FFLP Model

The arrival times $(\tilde{a})$, berthing times $(\tilde{m})$, and handling times $(\tilde{d})$ of the vessel are considered to be of fuzzy nature (imprecise)

In a similar way to the model of Sect. 5, the objective is to allocate all vessels according to several constraints minimizing the total waiting time, for all vessels.

Based on the deterministic model [7] and assuming the imprecision of some parameters and decision variables, we propose the following fuzzy model optimization.

$$\min \sum_{i \in V} (\tilde{m}_i - \tilde{a}_i) \tag{23}$$

Subject to:

$$\tilde{m}_i \geq \tilde{a}_i \quad \forall i \in V \tag{24}$$

$$p_i + l_i \leq L \quad \forall i \in V \tag{25}$$

$$p_i + l_i \leq p_j + M(1 - z_{ij}^x) \quad \forall i, j \in V, \ i \neq j \tag{26}$$

$$\tilde{m}_i + \tilde{h}_i \leq H \quad \forall i \in V \tag{27}$$

$$\tilde{m}_j + \tilde{h}_i + \ \leq \tilde{m}_i + M(1 - z_{ij}^y) \quad \forall i, j \in V, \ i \neq j \tag{28}$$

$$z_{ij}^x + z_{ji}^x + z_{ij}^y + z_{ji}^y \geq 1 \quad \forall i, j \in V, \ i \neq j \tag{29}$$

$$z_{ij}^x, z_{ij}^y \in \{0, 1\} \quad \forall i, j \in V, \ i \neq j. \tag{30}$$

The interpretation of constraints are similar to the model of Sect. 5, with the exception of the constraint (28). This constraint is regard to the time and indicate the vessel berths after or before another one.

6.2 Solution of the Model

We assume that all parameters and decision variables are linear and some of them are fuzzy. Thus, we have a fully fuzzy linear programming problem (FFLP).

The arrival of every vessel is represented by a triangular possibility distribution $\tilde{a} = (a_1, a_2, a_3)$, in a similar way, the berthing time is represented by $\tilde{m} = (m_1, m_2, m_3)$, and $\tilde{h} = (h_1, h_2, h_3)$ is considered a singleton.

When representing parameters and variables by triangular fuzzy numbers, we obtain a solution to the fuzzy model proposed applying the methodology proposed by Nasseri (see Sect. 3).

To apply this methodology, we use the operation of fuzzy difference on the objective function and the fuzzy sum on the constraints (see Sect. 2.2) and the First Index of Yagger as an ordering function on the objective function (see Sect. 2.3) obtaining the next auxiliary MILP model.

$$\min \sum_{i \in V} \frac{1}{3}((m1_i - a3_i) + (m2_i - a2_i) + (m3_i - a1_i)) \tag{31}$$

Subject to:

$$m1_i \geq a1_i \quad \forall i \in V \tag{32}$$

$$m2_i \geq a2_i \quad \forall i \in V \tag{33}$$

$$m3_i \geq a3_i \quad \forall i \in V \tag{34}$$

$$p_i + l_i \leq L \quad \forall i \in V \tag{35}$$

$$m3_i + h_i \leq H \quad \forall i \in V \tag{36}$$

$$p_i + l_i \leq p_j + M(1 - z_{ij}^x) \quad \forall i, j \in V, \ i \neq j \tag{37}$$

$$m1_i + h_i \leq m1_j + M(1 - z_{ij}^y) \quad \forall i, j \in V, \ i \neq j \tag{38}$$

$$m2_i + h_i \leq m2_j + M(1 - z_{ij}^y) \quad \forall i, j \in V, \ i \neq j \tag{39}$$

$$m3_i + h_i \leq m3_j + M(1 - z_{ij}^y) \quad \forall i, j \in V, \ i \neq j \tag{40}$$

$$m2_i > m1_i \quad \forall i \in V \tag{41}$$

$$m3_i > m2_i \quad \forall i \in V \tag{42}$$

$$z_{ij}^x + z_{ji}^x + z_{ij}^y + z_{ji}^y \geq 1 \quad \forall i, j \in V, \ i \neq j. \tag{43}$$

The planning horizon is the same that the model of Sect. 5.

Table 9 Fuzzy berthing plan obtained to the study case

Barcos	a1	a2	a3	m1	m2	m3	h	d1	d2	d3	l	p
V1	4	8	34	4	8	34	121	125	129	155	159	63
V2	0	15	36	0	15	36	231	231	246	267	150	222
V3	18	32	50	18	32	50	87	105	119	137	95	605
V4	9	40	46	9	40	46	248	257	288	294	63	0
V5	32	52	72	32	52	72	213	245	265	285	219	372
V6	55	68	86	245	265	285	496	741	761	781	274	332
V7	62	75	90	231	246	267	435	666	681	702	265	63
V8	45	86	87	105	119	137	146	251	265	283	94	606

6.3 Evaluacion

To the evaluation a personal computer equipped with a Core (TM) i5 - 4210U CPU 2.4 Ghz with 8.00 Gb RAM was used. The experiments was performed within a timeout of 60 min.

6.3.1 Evaluation of Study Case

To the vessels of study case (see Table 3), the berthing plan obtained with the model is showed in Table 9, and the polygonal-shaped are showed in Fig. 8.

The berthing plan showed in Table 9, is a fuzzy berthing one, e.g., to the vessel V6, the most possible arrival is at 68 units of time, but it could be early or late up to 55 and 86 units of time, respectively; the most possible berthing time is at 265 units of time, but it could berth between 245 and 285 units of time; the most possible departure time is at 761 units of time, but it could departure between 741 and 781 units of time.

An appropriate way to observe the robustness of the fuzzy berthing plan is the polygonal-shape representation (see Fig. 8). The line below the small triangle represents the possible early Berthing time; the line that is above the small triangle, the possible late time; the small triangle represents the optimum berthing time (with a greater possibility of occurrence) and the length of the polygon represents the time that vessel will stay at the quay.

In the circle of Fig. 8, we observe an apparent conflict between the departure time of vessel V2 and the berthing time of vessel V6. The conflict is not such, if the vessel V2 is late, the vessel V6 has slack times supporting delays. For example, assume that vessel V2 is late 10 units of time; according the Table 9, the berthing occurs at m = 15 + 10 = 25 units of time and its departure occurs at d = 25 + 231 = 256 units of time. The vessel V6 can moor during this space of time, since according to Table 9, its berthing can occurs between 245 and 285 units of time. This fact is observed in Fig. 9.

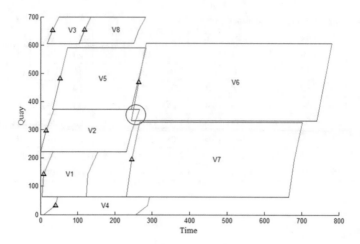

Fig. 8 Fuzzy berthing plan in polygonal-shape

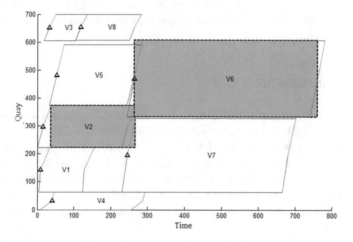

Fig. 9 Delayed berthing of vessel V2

In order to analyze the robustness of the fuzzy berthing plan, we simulate the incidences showed in Table 10.

With the incidences of Table 10, a feasible berthing plan can be obtained as showing in Table 11. In Fig. 10, we observe that the berthing plan obtained, is a part of the fuzzy plan obtained initially.

6.3.2   Evaluation of the Benchmark BAPQCAP-Imprecise

Table 12, shows the average of results obtained by CPLEX to the Benchmark BAPQCAP-Imprecise (see Sect. 4.2).

Table 10 Incidences in the vessel arrival times

Vessel	Time	Incidence
V1	13	Delay
V2	15	Delay
V3	0	On time
V4	18	Earliness
V5	10	Earliness
V6	8	Earliness
V7	9	Delay
V8	0	On time

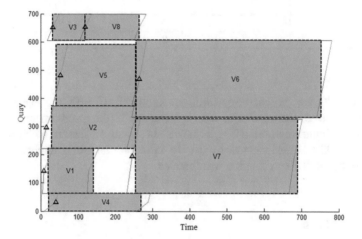

Fig. 10 Final berthing plan included in the fuzzy plan

Table 11 Final berthing plan including incidents

Barcos	Berthing time (m)	Handling time (h)	Departure time (d)	Length (l)	Position (p)
V1	21	121	142	159	63
V2	30	231	261	150	222
V3	32	87	119	95	605
V4	22	248	270	63	0
V5	42	213	255	219	372
V6	261	496	757	274	332
V7	261	435	696	265	63
V8	119	146	265	94	606

Table 12 Evaluation of imprecise benchmark

Vessels	Avs T	# Opt	# NOpt
5	859.65	100	0
10	5538.18	10	90
15	12729.66	0	100
20	23654.33	0	100
25	39631	0	100
30	59215.66	0	100
35	76323	0	100
40	93098.67	0	100
45	137863	0	100
50	–	0	0
55	–	0	0
60	–	0	0

From results, we can observe that in all cases solved by CPLEX, the objective function T increases as the number of vessels increases. To the given timeout, CPLEX, found the optimum solution in 90% of the instances with 10 vessels; a non-optimum solution in 100% of the instances from 15 to 45 vessels; and for a number of vessels greater or equal to 50 no solution was founded.

7 Conclusion

Even though many investigations about BAP have been carried out, most of them assume that vessel arrivals are deterministic. This is not real, in practice there are earliness or delays in vessel arrivals. Thus, the adaptability of a berthing plan is important for the global performance of the system in a MCT.

The results obtained show that the fuzzy MILP model to the BAP provides different berthing plans with different degrees of precision, but it also has an inconvenience: after the berthing time of a vessel, the next vessel has to wait all the time considered for the possible earliness and delay.

The model FFLP to the BAP surpass the inconvenient of the fuzzy MILP mode: in case the vessels arrive early or late a shorter time of the maximum tolerance, the fuzzy plan obtained can be adapted.

To a timeout of 60 min, both models allow to find the optimum solution for a small number of vessels, for instances from 15 up to 45 vessels they find non-optimum solutions and to greater number vessels they can't find no solutions.

The proposed models can be used when sufficient information is not available to obtain probability distributions on the arrival time of vessels that will allow posing a stochastic model.

Likewise they can be used when we want to find berthing plans on the basis of inaccurate information obtained in advance about the vessel arrivals. For every vessel it is necessary to request the time interval of possible arrival, as well as the more possible time the arrival occurs.

Finally, because of this research, we have open problems for future researches: To extend the model that considers the quay cranes to be assigned to every vessel. To use meta-heuristics to solve the fuzzy BAP model more efficiently, when the number of vessels is greater.

Acknowledgements This work was supported by INNOVATE-PERU, Project N PIBA-2-P-069-14.

References

1. Bierwirth, C., Meisel, F.: A survey of berth allocation and quay crane scheduling problems in container terminals. Eur. J. Oper. Res. **202**(3), 615–627 (2010)
2. Bruggeling, M., Verbraeck, A., Honig, H.: Decision support for container terminal berth planning: integration and visualization of terminal information. In: Proceedings of the Van de Vervoers logistieke Werkdagen (VLW2011), University Press, Zelzate, pp. 263–283 (2011)
3. Das, S.K., Mandal, T., Edalatpanah, S.A.: A mathematical model for solving fully fuzzy linear programming problem with trapezoidal fuzzy numbers. Appl. Intell. **46**(3), 509–519 (2017)
4. Exposito-Izquiero, C., Lalla-Ruiz, E., Lamata, T., Melian-Batista, B., Moreno-Vega, J.: Fuzzy optimization models for seaside port logistics: berthing and quay crane scheduling. In: Computational Intelligence, pp. 323–343. Springer International Publishing (2016)
5. Gutierrez, F., Vergara, E., Rodrguez, M., Barber, F.: Un modelo de optimizacin difuso para el problema de atraque de barcos. Investigacin Operacional **38**(2), 160–169 (2017)
6. Jimenez, M., Arenas, M., Bilbao, A., Rodrı, M.V.: Linear programming with fuzzy parameters: an interactive method resolution. Eur. J. Oper. Res. **177**(3), 1599–1609 (2007)
7. Kim, K.H., Moon, K.C.: Berth scheduling by simulated annealing. Transp. Res. Part B Methodol. **37**(6), 541–560 (2003)
8. Laumanns, M., et al.: Robust adaptive resource allocation in container terminals. In: Proceedings of the 25th Mini-EURO Conference Uncertainty and Robustness in Planning and Decision Making, Coimbra, Portugal, pp. 501–517 (2010)
9. Lim, A.: The berth planning problem. Oper. Res. Lett. **22**(2), 105–110 (1998)
10. Luhandjula, M.K.: Fuzzy mathematical programming: theory, applications and extension. J. Uncertain Syst. **1**(2), 124–136 (2007)
11. Nasseri, S.H., Behmanesh, E., Taleshian, F., Abdolalipoor, M., Taghi-Nezhad, N.A.: Fullyfuzzy linear programming with inequality constraints. Int. J. Ind. Math. **5**(4), 309–316 (2013)
12. Rodriguez-Molins, M., Ingolotti, L., Barber, F., Salido, M.A., Sierra, M.R., Puente, J.: A genetic algorithm for robust berth allocation and quay crane assignment. Prog. Artif. Intell. **2**(4), 177–192 (2014)
13. Rodriguez-Molins, M., Salido, M.A., Barber, F.: A GRASP-based metaheuristic for the Berth Allocation Problem and the Quay Crane Assignment Problem by managing vessel cargo holds. Appl. Intell. **40**(2), 273–290 (2014)
14. Steenken, D., Vo, S., Stahlbock, R.: Container terminal operation and operations research—a classification and literature review. OR Spectr. **26**(1), 3–49 (2004)
15. Wang, X., Kerre, E.E.: Reasonable properties for the ordering of fuzzy quantities (I). Fuzzy Sets Syst. **118**(3), 375–385 (2001)
16. Yager, R.R.: A procedure for ordering fuzzy subsets of the unit interval. Inf. Sci. **24**(2), 143–161 (1981)

17. Young-Jou, L., Hwang, C.: Fuzzy Mathematical Programming: Methods and Applications, vol. 394. Springer Science & Business Media (2012)
18. Zadeh, L.A.: Fuzzy sets as a basis for a theory of possibility. Fuzzy Sets Syst. **100**, 9–34 (1999)
19. Zimmermann, H.: Fuzzy Set Theory and Its Applications, Fourth Revised Edition. Springer, Berlin (2001)

Identifying Clusters in Spatial Data Via Sequential Importance Sampling

Nishanthi Raveendran and Georgy Sofronov

Abstract Spatial clustering is an important component of spatial data analysis which aims in identifying the boundaries of domains and their number. It is commonly used in disease surveillance, spatial epidemiology, population genetics, landscape ecology, crime analysis and many other fields. In this paper, we focus on identifying homogeneous sub-regions in binary data, which indicate the presence or absence of a certain plant species which are observed over a two-dimensional lattice. To solve this clustering problem we propose to use the change-point methodology. we consider a Sequential Importance Sampling approach to change-point methodology using Monte Carlo simulation to find estimates of change-points as well as parameters on each domain. Numerical experiments illustrate the effectiveness of the approach. We applied this method to artificially generated data set and compared with the results obtained via binary segmentation procedure. We also provide example with real data set to illustrate the usefulness of this method.

1 Introduction

Identifying homogeneous domains is of particular interest in spatial statistics. It is usually the case that spatial data have pre-defined subdivisions of interest. For example, data are often collected on non-overlapping administrative or census districts and these districts may have irregular shapes; see [27]. Spatial clustering is an important component of statistical analysis since spatial data may be heterogeneous and it may be difficult to interpret the parameters of the corresponding statistical model. However, if we cluster the data into homogeneous domains, then we can construct appropriate statistical models for each cluster. The problem of finding regional homogeneous domains is known as segmentation, partitioning or clustering. The two

N. Raveendran · G. Sofronov (✉)
Department of Statistics, Macquarie University, Sydney, NSW 2109, Australia
e-mail: georgy.sofronov@mq.edu.au

N. Raveendran
e-mail: nishanthi.raveendran@mq.edu.au

© Springer Nature Switzerland AG 2019
S. Fidanova (ed.), *Recent Advances in Computational Optimization*,
Studies in Computational Intelligence 795,
https://doi.org/10.1007/978-3-319-99648-6_10

main problems in spatial clustering are identifying the number of domains, which is usually not known in advance, and estimating the boundaries of such domains.

Many clustering algorithms have been developed in the literature, ranging from hierarchical methods such as bottom-up (or agglomerative) methods and top-down (or divisive) methods, to optimization methods such as the k-means algorithm [5]. The algorithms have numerous applications in pattern recognition, spatial data analysis, image processing, market research; see [24]. Spatial clustering covers enormous practical problems in many disciplines. For example, in epidemiological studies and public health research, it is known that the disease risk varies across space and it is important to identify regions of safety and regions of risk. A model using Bayesian approach for spatial clustering was discussed in [10]. Recently, a two-stage Bayesian approach for estimating the spatial pattern in disease risk and identifying clusters which have high (or low) disease risks was proposed in [1].

The homogeneity changes in space is an important research subject in ecology. In a large area, the spatial distribution of plant or animal species is very unlikely to be homogeneous. Studying these kinds of changes is important in several ways. For example, detecting early changes in vegetation may improve productivity. A class of Bayesian statistical models to identify thresholds and their locations in ecological data was introduced in [2]. A method to estimate the change-point distribution between two patches was presented in [13].

Studies of weather and climatic systems at a global scale have become a prime area of research for a number of reasons; one of these is the concern about global climatic change. Mann-Kendall trend test, Bayesian change-point analysis and a hidden Markov model to find changes in the rainfall and temperature patterns over India are used in [23].

There has also been extensive literature on image recognition with some articles presenting statistical approaches to the boundary identification in statistical imaging. For example, [11] presented a Markov chain Monte Carlo (MCMC) method to identify closed object boundaries in gray-scale images. Change curve estimation problem is also referred as multidimensional detection problem or boundary estimation problem. A wavelet method to estimate jumps and sharp curves in the plane was proposed in [26].

In this study, we are interested in identifying the boundaries of spatial domains with applications to an ecological landscape. In general, these problems are typically challenging due to the multivariate nature of the data which leads to complex and highly parameterised likelihoods. We use binary data indicating the presence or absence of plant species, which are observed over a two-dimensional lattice. We consider our problem as a change-point detection problem, which is commonly used in analysing time series to detect changes and their locations. For more on change-point methods and applications, see [4, 8, 17]. In this paper, we present a Sequential Importance Sampling (SIS) approach to change-point modelling using Monte Carlo simulation to find estimates of spatial domains as well as parameters of the process on each domain. We include results of numerical experiments indicating the usefulness of this method.

Binary spatial data are commonly involved in various areas such as economics, social sciences, ecology, image analysis and epidemiology. Also, such data frequently occur in environmental and ecological research, for instance, when the data correspond to presence or absence of a certain invasive plant species at a location or, when the data happen to fall into one of two categories, say, two types of soil. The general overview of spatial data can be found in [6, 7, 25].

This study aims to develop effective procedures based on SIS procedure for estimating homogeneous domains and their boundaries in spatial data. This paper is organized as follows. Section 2 introduces the spatial segmentation problem. In Sect. 3, we introduce the essential concepts of SIS. In Sect. 4, we develop SIS for the spatial segmentation problem. In Sect. 5, we consider two examples illustrating how well SIS can work for generated and real data. Section 6 provides concluding remarks.

2 The Spatial Segmentation Problem

Let us formally introduce the spatial segmentation problem. We are given spatial binary data, a lattice with zeros and ones (presence-absence data), which can be represented by an $m_1 \times m_2$ matrix, B. A segmentation of B is specified by giving the number of domains N and the domains $D = (D_1, \ldots, D_N)$ with their boundaries parallel to the sides of the lattice. The position of each domain $D_n, n = 1, \ldots, N$, can be defined by the four edges or cuts $c_n = (c_{n1}, c_{n2}, c_{n3}, c_{n4}) = (c_{nL}, c_{nR}, c_{nU}, c_{nD})$, left, right, up and down, respectively. A maximum number of domains $N_{\max}$ is specified, where $1 \leq N \leq N_{\max} < m_1 m_2$. The model for the data assumes that within each domain characters are generated by independent Bernoulli trials with probability of success (presence of the plant indicates "1") θ_n that depends on the domain. Thus, the joint probability density function of spatial data B conditional on N, $D = (D_1, \ldots, D_N)$, and $\theta = (\theta_1, \ldots, \theta_N)$, is given by

$$f(B \mid N, D, \theta) = \prod_{n=1}^{N} \theta_n^{I_{D_n}} (1 - \theta_n)^{O_{D_n}},$$

where I_{D_n} is the number of ones in domain D_n and O_{D_n} is the number of zeros in that same domain.

To use the Bayesian framework, a prior distribution must be defined on the set of possible values of $\mathbf{x} = (N, D, \theta)$, denoted

$$\mathscr{X} = \cup_{N=1}^{N_{\max}} \{N\} \times \mathscr{C}_N \times (0, 1)^N$$

with

$$\mathscr{C}_N = \{(c_1, \ldots, c_N) : c_n = (c_{n1}, c_{n2}, c_{n3}, c_{n4}), c_{n1} < c_{n2}, c_{n3} < c_{n4},$$
$$c_{n1} \in \{0, \ldots, m_2 - 1\}, c_{n2} \in \{1, \ldots, m_2\},$$
$$c_{n3} \in \{0, \ldots, m_1 - 1\}, c_{n4} \in \{1, \ldots, m_1\}\},$$

where $c_{n1} = 0$, $c_{n2} = m_2$, $c_{n3} = 0$, or $c_{n4} = m_1$ means that the boundary of the n-th domain coincides with the edge of the lattice.

We assume a uniform prior both on the number of domains and on $\mathscr{C}_N$, and uniform priors on $(0, 1)$ for each θ_n. This means that the overall prior $f_0(N, D, \theta)$ is a constant. Therefore, the posterior density function at point $\mathbf{x} = (N, D, \theta)$, having observed spatial data B, is given by

$$f(\mathbf{x} \mid B) = \pi(\mathbf{x}) \propto \prod_{n=1}^{N} \theta_n^{I_{D_n}} (1 - \theta_n)^{O_{D_n}}.$$

3   Sequential Importance Sampling

Consider the problem where we wish to evaluate the quantity

$$\ell = \int_{\mathscr{X}} G(\mathbf{x}) \pi(\mathbf{x}) \, d\mathbf{x} = \mathbb{E}_\pi[G(\mathbf{X})],$$

where the subscript π means that the expectation is taken with respect to $\pi(\mathbf{x})$ — the target density (in our case the posterior density) — and $G(\mathbf{x}) \geq 0$ is some performance function.

We can then represent ℓ as:

$$\ell = \int G(\mathbf{x}) \frac{\pi(\mathbf{x})}{q(\mathbf{x})} q(\mathbf{x}) \, d\mathbf{x} = E_q \left[G(\mathbf{X}) \frac{\pi(\mathbf{X})}{q(\mathbf{X})} \right].$$

We can now get an unbiased estimator for ℓ, called the *importance sampling estimator*, as follows:

$$\hat{\ell} = \frac{1}{M} \sum_{i=1}^{M} G(\mathbf{X}^{(i)}) \frac{\pi(\mathbf{X}^{(i)})}{q(\mathbf{X}^{(i)})},$$

where $\mathbf{X}^{(1)}, \ldots, \mathbf{X}^{(M)}$ is a random sample from a different density q. The ratio of densities

$$W(\mathbf{x}) = \frac{\pi(\mathbf{x})}{q(\mathbf{x})}$$

is called the *importance weight* or *likelihood ratio*.

A variant of the importance sampling technique is known as *sequential importance sampling* (SIS). It is not always easy to come up with an appropriately close importance sampling density $q(\mathbf{x})$ for high-dimensional target distributions. SIS builds up the importance sampling density sequentially.

Suppose that $\mathbf{x}$ can be written in the form $\mathbf{x} = (x_1, x_2, \ldots, x_d)$, where each of the x_i may be multi-dimensional. Then we may construct our importance sampling density as

$$q(\mathbf{x}) = q_1(x_1)\, q_2(x_2 \mid x_1) \cdots q_d(x_d \mid x_1, \ldots, x_{d-1}),$$

where the q_t are chosen so as to make $q(\mathbf{x})$ as close to the target density, $\pi(\mathbf{x})$, as possible. We can also rewrite the target density sequentially as

$$\pi(\mathbf{x}) = \pi(x_1)\pi(x_2 \mid x_1) \cdots \pi(x_d \mid \mathbf{x}_{d-1}),$$

where we have abbreviated $(x_1, \ldots, x_t)$ to $\mathbf{x}_t$. The likelihood ratio now becomes

$$w_d = \frac{\pi(x_1)\pi(x_2 \mid x_1) \cdots \pi(x_d \mid \mathbf{x}_{d-1})}{q_1(x_1)\, q_2(x_2 \mid x_1) \cdots q_d(x_d \mid \mathbf{x}_{d-1})},$$

which can be evaluated sequentially as

$$w_t = u_t\, w_{t-1}, \quad t = 1, \ldots, d,$$

with initial weight $w_0 = 1$. The *incremental weights* $\{u_t\}$ are given by $u_1 = \pi(x_1)/q_1(x_1)$ and

$$u_t = \frac{\pi(x_t \mid \mathbf{x}_{t-1})}{q_t(x_t \mid \mathbf{x}_{t-1})} = \frac{\pi(\mathbf{x}_t)}{\pi(\mathbf{x}_{t-1})\, q_t(x_t \mid \mathbf{x}_{t-1})},$$
$$t = 2, \ldots, d.$$

However this incremental weight requires knowing the marginal probability density functions $\{\pi(\mathbf{x}_t)\}$. This may not be easy and so we need to introduce a sequence of *auxiliary* probability density functions $\pi_1, \pi_2, \ldots, \pi_d$ such that (a) $\pi_t(\mathbf{x}_t)$ is a good approximation to $\pi(\mathbf{x}_t)$, (b) they are easy to evaluate, and (c) $\pi_d = \pi$. The SIS method can now be described as follows.

Algorithm 1 (*Sequential Importance Sampling*)

1. For each finite $t = 1, \ldots, d$, draw $X_t = x_t$ from $q(x_t \mid \mathbf{x}_{t-1})$.
2. Compute $w_t = u_t w_{t-1}$, where $w_0 = 1$ and

$$u_t = \frac{\pi_t(\mathbf{x}_t)}{\pi_{t-1}(\mathbf{x}_{t-1})\, q_t(\mathbf{x}_t \mid \mathbf{x}_{t-1})},$$
$$t = 1, \ldots, d.$$

3. Repeat M times and estimate ℓ via

$$\hat{\ell}_w = \frac{w^{(1)}G(\mathbf{x}^{(1)}) + \cdots + w^{(M)}G(\mathbf{x}^{(M)})}{w^{(1)} + w^{(2)} + \cdots + w^{(M)}},$$

with $w^{(i)} \equiv w_d^{(i)}$ for $i = 1, \ldots, M$.

For more on importance sampling and SIS see, [19, 21].

4   SIS for the Spatial Segmentation Problem

In order to identify homogeneous domains, we consider the evaluation of the following integral:

$$\ell(y) = \sum_{N=1}^{N_{\max}} \sum_{D \in \mathscr{C}_N} \int_{(0,1)^N} G(y)\pi(N, D, \theta)\, d\theta,$$

where

$$G(y) = \theta_n \quad \text{if} \quad y = (y_1, y_2) \in D_n.$$

The change-point variable $\mathbf{x}$ can be represented as a d-dimensional vector, $d = N_{\max} - 1$:

$$\mathbf{x} = (x_1, \ldots, x_d), \quad x_t = (c_{jt}', \theta_{0,t}', \theta_{1,t}'), \quad j = 0, 1, \quad 1 \le t \le d,$$

where

- c_{jt}' is the position of a cut (either a column index, $j = 0$, or a row index, $j = 1$), which has been defined at the t-th iteration of Algorithm 2;
- $\theta_{0,t}'$ and $\theta_{1,t}'$ are values of the parameter for the two new domains obtained at the t-th iteration, $D_{0,t}$ and $D_{1,t}$ (see, for example, Fig. 1, part (a)).

We will define $\theta_{0,t}' = \theta_{1,t}'$, if there is no cut at the t-th iteration. If at each iteration we obtain a new cut, then we have $N_{\max}$ domains, $D_1, \ldots, D_{N_{\max}}$. If at some iterations there is no cuts, then we obtain N, $N < N_{\max}$, domains, $D_1, \ldots, D_N$.

In order to use Algorithm 1, we need to construct the importance sampling (or proposal) density $q(\mathbf{x})$. Using the initial data $D^{(1)} = B$, we generate a point $x_1 = (c_{j1}', \theta_{0,1}', \theta_{1,1}')$ by simulation from a distribution

$$q_1(x_1) \propto f(B \mid x_1)f_1(x_1),$$

where $f_1(x_1)$ is a prior density defined on $\{1, \ldots, m_1 + m_2\} \times (0, 1)^2$. Since we have assumed a uniform prior on $\mathscr{C}_N$ and uniform priors on $(0, 1)$ for each θ_n, then $f_1(x_1)$ is a constant.

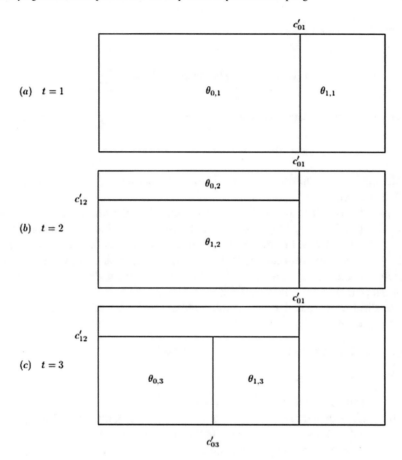

Fig. 1 Three iterations of the SIS procedure

Now we have two domains $D_{0,1}$ and $D_{1,1}$ to the up and down of c'_{11} or the left and right of c'_{01} (one of the domains may be empty). We choose these domains proportional to their size, $S_{0,1}$ and $S_{1,1}$, respectively. Since $D_{0,1}$ is independent of $D_{1,1}$, there is no need to know the values of the parameter θ for non-selected domains. Using the selected domain $D^{(2)} \in \{D_{0,1}, D_{1,1}\}$, we generate the next point $x_2 = (c'_{j2}, \theta'_{0,2}, \theta'_{1,2})$ such that c'_{j2} cuts domain $D^{(2)}$ in two parts. Then we obtain

$$q_2(x_2 \mid x_1) \propto v^{(2)} f(D^{(2)} \mid x_2),$$

where $v^{(2)} \in \{v_{0,1}, v_{1,1}\}$, $v_{k,1} = S_{k,1}/(S_{0,1} + S_{1,1})$.

Following this process, at the t-th iteration we have

$$q_t(x_t \mid \mathbf{x}_{t-1}) \propto v^{(t)} f(D^{(t)} \mid x_t), \quad t = 2, \ldots, d,$$

where

$$v^{(t)} \in \{v_{0,t-1}, \ldots, v_{t-1,t-1}\}, \quad v_{k,t-1} = S_{k,t-1} \left(\sum_{k=0}^{t-1} S_{k,t-1} \right)^{-1}.$$

We can define the sequence of auxiliary probability density functions $\pi_1, \ldots, \pi_d$ as

$$\pi_t(\mathbf{x}_t) \propto f(B \mid \mathbf{x}_t) f_t(\mathbf{x}_t), \quad t = 1, \ldots, d,$$

where $f_t(\mathbf{x}_t)$ is a uniform prior distribution. In particular, $\pi_1(x_1) = q_1(x_1)$ and $\pi_d(\mathbf{x}_d) = \pi(\mathbf{x})$.

Figure 1 shows the first three iterations of the SIS algorithm for the spatial segmentation problem. In part (a), we have a domain and we find our first column-cut $(j = 0)$ indicated by the line c'_{01}. We now draw $\theta_{0,1}$ and $\theta_{1,1}$ for the domains to the left and right of this cut, respectively. In the second iteration (part (b)), we have picked the left domain proportional to its size. We draw a row-cut c'_{12} with $\theta_{0,2}$ and $\theta_{1,2}$. This is repeated until we have drawn d cuts.

The SIS for the spatial segmentation problem can be described as the following iterative procedure.

Algorithm 2 (*SIS for Spatial Segmentation Problem*)

1. Draw $X_t = x_t$ from $q_t(x_t \mid \mathbf{x}_{t-1})$. That is,

   a. Calculate the weights $v_{k,t-1}, k = 0, \ldots, t-1$.
   b. Select a domain $D^{(t)} \in \{D_{k,t-1}, k = 0, \ldots, t-1\}$ with probabilities proportional to the weights calculated in the previous step.
   c. Calculate the posterior probabilities

   $$f(c'_{jt} \mid D^{(t)}) \propto p_1(c'_{jt}) \int_0^1 \theta^{I_{D_{0,t}}} (1-\theta)^{O_{D_{0,t}}} d\theta \int_0^1 \theta^{I_{D_{1,t}}} (1-\theta)^{O_{D_{1,t}}} d\theta,$$

   where $p_1(c'_{jt})$ is a uniform prior distribution.
   d. Insert a new cut at c'_{jt} proportional to the probabilities calculated in the previous step.
   e. Select new Bernoulli parameters $\theta'_{0,t}$ and $\theta'_{1,t}$ for the two new domains by sampling from the Beta distribution with parameters $(\alpha_{0,t} = I_{D_{0,t}} + 1, \beta_{0,t} = O_{D_{0,t}} + 1)$ and $(\alpha_{1,t} = I_{D_{1,t}} + 1, \beta_{1,t} = O_{D_{1,t}} + 1)$, respectively.

   Let $\mathbf{x}_t = (\mathbf{x}_{t-1}, x_t)$, where $x_t = (c'_{jt}, \theta'_{0,t}, \theta'_{1,t})$.
2. Compute

   $$u_t = \frac{\pi_t(\mathbf{x}_t)}{\pi_{t-1}(\mathbf{x}_{t-1}) q_t(x_t \mid \mathbf{x}_{t-1})}$$

   and let $w_t = w_{t-1} u_t, w_0 = 1, t = 1, \ldots, d$.

3. Repeat M times and estimate ℓ via

$$\hat{\ell}_w = \frac{w^{(1)} G(\mathbf{x}^{(1)}) + \cdots + w^{(M)} G(\mathbf{x}^{(M)})}{w^{(1)} + w^{(2)} + \cdots + w^{(M)}},$$

with $w^{(i)} \equiv w_d^{(i)}$ for $i = 1, \ldots, M$.

5 Results

In this section, we discuss two examples with artificially generated and real spatial data sets to illustrate the usefulness of the proposed algorithm. We compare our SIS approach to binary segmentation (BS) algorithm, which is developed for the same spatial problem [18]. The first data set is an artificial matrix with a known distribution, while in the second example we use real presence-absence plant data [22].

5.1 Binary Segmentation Algorithm

In order to identify homogeneous domains and their boundaries in spatial data, we developed new methods that are based on binary segmentation algorithm which is a well-known multiple change-point detection method. This method is effective in identifying multiple domains and their boundaries in two dimensional spatial data. The benefits of binary segmentation include low computational (typically of order $O(n)$), conceptual simplicity, the fact that it is usually easy to code, even in more complex models, and at each stage it involves one-dimensional rather than multi-dimensional optimization. On the other hand, the method is "greedy" procedure in the sense that it is performed sequentially, with each stage depending on the previous ones, which are never re-visited. For more on binary segmentation methods and applications, see [3, 9, 12, 15, 27].

This algorithm uses the maximum likelihood test for a single change-point detection problem. It searches every column and row to detect an optimal cut, which maximises the test statistic. If the test statistic greater than the threshold value, it splits the domain according to the index (row or column) and stores the obtained domains. Otherwise, the algorithm stops. This procedure is repeated until a stopping criterion is met. In general, this method can be summarized by a three-step iterative procedure.

1. Given the data, search the change point column-wise and find the optimal cut which maximizes the test statistic. Repeat this procedure row-wise.
2. Select the maximum of the two test statistics for the optimal column and row cuts and compare with the threshold value. If the test statistic is greater than the threshold value, then split the data in two domains.
3. Repeat steps 1 and 2 for each domain until no new domains are identified.

5.2   Example 1: Artificial Data

Let B be a 30×30 matrix of independent Bernoulli random variables generated with
the parameters given in Table 1. The true profile of this matrix can be seen in Fig. 2.

We used the SIS algorithm in Algorithm 2 with $d = 5$, and, for comparison,
$M = 100$ and 500. We use the parameter of Bernoulli distribution to calculate the
Root Mean Squared Error (RMSE). We consider the values for each cell of the matrix
B. The estimated values for each cell within a particular obtained domain remains
the same. The RMSE is calculated as

$$\text{RMSE} = \sqrt{\frac{\sum_{i=1}^{m_1} \sum_{j=1}^{m_2} (e_{ij} - t_{ij})^2}{m_1 m_2}},$$

Table 1   The parameters of the generated data matrix

Domains	Coordinates (top left to bottom right)	Probability (p_i)
Domain 1	(1,1)–(20,10)	$p_1 = 0.35$
Domain 2	(1,10)–(20,30)	$p_2 = 0.40$
Domain 3	(20,1)–(30,10)	$p_3 = 0.40$
Domain 4	(20,10)–(30,30)	$p_4 = 0.45$

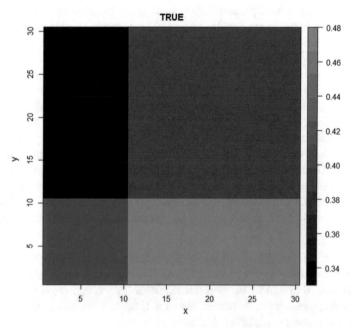

Fig. 2   Four domains with different parameters of Bernoulli distribution shown by different colour

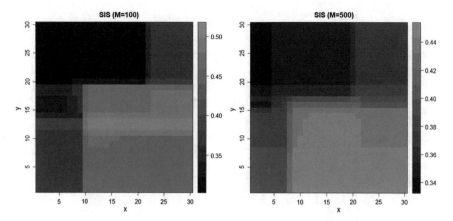

Fig. 3 The profiles estimated by the SIS algorithm with $M = 100$ (left) and $M = 500$ (right)

where e_{ij} and t_{ij} are the estimated and the true values, respectively, for cell (i, j) of the matrix (i, j are the corresponding row and column); $m_1 \times m_2$ is the size of the matrix.

Figure 3 shows the true profile and obtained domains from SIS approach. The two plots are in excellent agreement with true profile, supporting the fact that both produced only very small difference between their estimates and the true distribution. It is interesting to note that our algorithm almost always under-estimated or over-estimated the boundary of the domains. This could be attributed to the fact that although the matrix was drawn from Bernoulli distributions with parameters given in Table 1, it is still artificially generated data and it is possible that the true profile may not be exactly the same as the Bernoulli parameters of Table 1. It is also clear from Fig. 3 that the estimates become more accurate when M increases.

Figure 4 shows the obtained domains from both SIS and BS algorithms for the same example in Table 1. These two plots are in excellent agreement with each other and the RMSE values for SIS and BS are 0.0345 and 0.0530. In [18], we show that BS performs well in identifying the correct position of the cut when the differences between the parameters are rather high.

5.3 Example 2: Real Data

The second example uses a real plant data [22]. The authors study *Alchemilla* plants and their microspecies growing within Eastern Europe with particular interest in the spatial distribution of the microspecies. The plants were considered on square sites with the area of 1 m². The surveyed area in locality M1 was 169 m².

Figure 5 shows the real data, where a black cell indicates the presence of the plant, while a white cell means its absence. So we can consider a matrix encoded

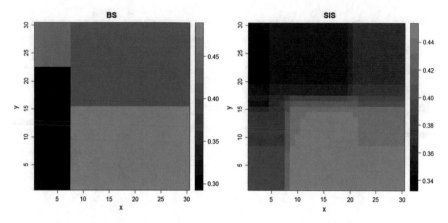

Fig. 4 Domains determined by the BS (left) algorithm and the profile estimated by SIS (right) algorithm

101	102	103	104	105	106	107	108	109	110	201	211	221
111	112	113	114	115	116	117	118	119	120	202	212	222
121	122	123	124	125	126	127	128	129	130	203	213	223
131	132	133	134	135	136	137	138	139	140	204	214	224
141	142	143	144	145	146	147	148	149	150	205	215	225
151	152	153	154	155	156	157	158	159	160	206	216	226
161	162	163	164	165	166	167	168	169	170	207	217	227
171	172	173	174	175	176	177	178	179	180	208	218	228
181	182	183	184	185	186	187	188	189	190	209	219	229
191	192	193	194	195	196	197	198	199	200	210	220	230
231	232	233	234	235	236	237	238	239	240	261	262	263
241	242	243	244	245	246	247	248	249	250	264	265	266
251	252	253	254	255	256	257	258	259	260	267	268	269

Fig. 5 Layout of sites in locality M1. The 169 sites are numbered from 101 to 269. If a site has microspecies *A. tubulosa* Juz. then it is colour coded black, otherwise it is white

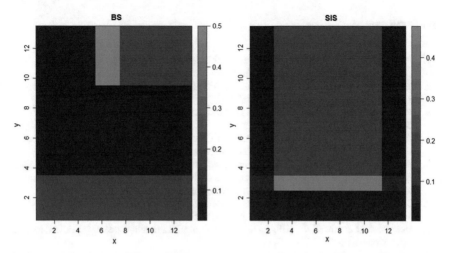

Fig. 6 Domains determined by the BS (left) and SIS (right) algorithms applied to the real data

with zeros (white) and ones (black). As a consequence of studying a real data set, we do not know its true profile but we can still look for agreement between two algorithms (BS and SIS). Using the same parameters as in the previous example, we obtain the results summarized in Fig. 6. It shows that the obtained domains from SIS and BS methods differ from each other. The segmentation obtained by SIS indicates that the left, bottom and right edges of the studied area are quite different compare to its middle part, which has a higher density of plants. One of the reasons is that the right and bottom clusters are somewhat concordant with the area that was added at a later stage of the experiment (the sites numbered from 201 to 269), which may have affected the results of the study.

6 Conclusion

In this paper we have developed a new Sequential Importance Sampling algorithm to identify homogeneous domains in spatial data. We have demonstrated the effectiveness of this method in examples using both real and artificial data sets. For the artificial data set, domains obtained by the SIS were in excellent agreement with both the Binary Segmentation (BS) and the true profile. When comparing these two methods to real data, the SIS approach produces the result that was clearer for interpretation. Both algorithms work well and, while the BS is a deterministic algorithm, which always produces the same result, the SIS is based on Monte Carlo simulations. We proposed and compared two spatial segmentation algorithms, the development of new spatial segmentations algorithms based on other well-known statistical computational methods such as Cross Entropy (CE) [16] and Markov chain Monte Carlo (MCMC) [14, 20] methods is a matter of our further research.

References

1. Anderson, C., Lee, D., Dean, N.: Bayesian cluster detection via adjacency modelling. Spat. Spatio-Temporal Epidemiol. **16**, 11–20 (2016)
2. Beckage, B., Joseph, L., Belisle, P., Wolfson, D.B., Platt, W.J.: Bayesian change-point analyses in ecology. New Phytol. **174**(2), 456–467 (2007)
3. Braun, J.V., Muller, H.-G.: Statistical methods for DNA sequence segmentation. Stat. Sci. **13**(2), 142–162 (1998)
4. Chen, J., Gupta, A.K.: Parametric Statistical Change Point Analysis: With Applications to Genetics, Medicine, and Finance. Springer Science & Business Media (2011)
5. Chen, S.S., Gopalakrishnan, P.S.: Clustering via the bayesian information criterion with applications in speech recognition. In: Proceedings of the 1998 IEEE International Conference on Acoustics, Speech and Signal Processing, 1998, vol. 2, pp. 645–648. IEEE (1998)
6. Cliff, A.D., Ord, J.K.: Spatial Processes: Models & Applications. Taylor & Francis (1981)
7. Cressie, N.: Statistics for Spatial Data. Wiley (2015)
8. Eckley, I.A., Fearnhead, P., Killick, R.: Analysis of changepoint models. In: Bayesian Time Series Models, pp. 205–224 (2011)
9. Fryzlewicz, P., et al.: Wild binary segmentation for multiple change-point detection. Ann. Stat. **42**(6), 2243–2281 (2014)
10. Gangnon, R.E., Clayton, M.K.: Bayesian detection and modeling of spatial disease clustering. Biometrics **56**(3), 922–935 (2000)
11. Helterbrand, J.D., Cressie, N., Davidson, J.L.: A statistical approach to identifying closed object boundaries in images. Adv. Appl. Probab. **26**(04), 831–854 (1994)
12. Killick, R., Eckley, I.A., Jonathan, P., Chester, U.: Efficient detection of multiple changepoints within an oceano-graphic time series. In: Proceedings of the 58th World Science Congress of ISI (2011)
13. López, I., Gámez, M., Garay, J., Standovár, T., Varga, Z.: Application of change-point problem to the detection of plant patches. Acta Biotheor. **58**(1), 51–63 (2010)
14. Manuguerra, M., Sofronov, G., Tani, M., Heller, G.: Monte Carlo methods in spatio-temporal regression modeling of migration in the EU. In: 2013 IEEE Conference on Computational Intelligence for Financial Engineering & Economics (CIFEr), pp. 128–134. IEEE (2013)
15. Olshen, A.B., Venkatraman, E., Lucito, R., Wigler, M.: Circular binary segmentation for the analysis of array-based DNA copy number data. Biostatistics **5**(4), 557–572 (2004)
16. Polushina, T., Sofronov, G.: Change-point detection in binary Markov DNA sequences by the cross-entropy method. In: 2014 Federated Conference on Computer Science and Information Systems (FedCSIS), pp. 471–478. IEEE (2014)
17. Priyadarshana, W., Sofronov, G.: Multiple break-points detection in array CGH data via the cross-entropy method. IEEE/ACM Trans. Comput. Biol. Bioinf. (TCBB) **12**(2), 487–498 (2015)
18. Raveendran, N., Sofronov, G.: Binary segmentation methods for identifying boundaries of spatial domains. Computer Science and Information Systems (FedCSIS) **13**, 95–102 (2017)
19. Rubinstein, R.Y., Kroese, D.P.: Simulation and the Monte Carlo Method, vol. 10, 2nd edn. Wiley (2007)
20. Sofronov, G.: Change-point modelling in biological sequences via the bayesian adaptive independent sampler. Int. Proc. Comput. Sci. Inf. Technol. **5**(2011), 122–126 (2011)
21. Sofronov, G.Y., Evans, G.E., Keith, J.M., Kroese, D.P.: Identifying change-points in biological sequences via sequential importance sampling. Environ. Model. & Assess. **14**(5), 577–584 (2009)
22. Sofronov, G.Y., Glotov, N.V., Zhukova, O.V.: Statistical analysis of spatial distribution in populations of microspecies of Alchemilla l. In: Proceedings of the 30th International Workshop on Statistical Modelling, vol. 2, pp. 259–262. Statistical Modelling Society, Linz, Austria (2015)
23. Tripathi, S., Govindaraju, R.S.: Change detection in rainfall and temperature patterns over India. In: Proceedings of the Third International Workshop on Knowledge Discovery from Sensor Data, pp. 133–141. ACM (2009)

24. Tung, A.K., Hou, J., Han, J.: Spatial clustering in the presence of obstacles. In: Proceedings of 17th International Conference on Data Engineering, 2001, pp. 359–367. IEEE (2001)
25. Upton, G.J., Fingleton, B.: Spatial Data Analysis by Example, vol. 1: Point Pattern and Quantitative Data. Wiley, Chichester, 1 (1985)
26. Wang, Y.: Change curve estimation via wavelets. J. Am. Stat. Assoc. **93**(441), 163–172 (1998)
27. Yang, T.Y., Swartz, T.B.: Applications of binary segmentation to the estimation of quantal response curves and spatial intensity. Biometrical J. **47**(4), 489–501 (2005)

Multiobjective Optimization Grover Adaptive Search

Benjamín Barán and Marcos Villagra

Abstract Quantum computing is a fast evolving subject with a promise of highly efficient solutions to difficult and complex problems in science and engineering. With the advent of large-scale quantum computation, a lot of effort is invested in finding new applications of quantum algorithms. In this article, we propose an algorithm based on Grover's adaptative search for multiobjective optimization problems where access to the objective functions is given via two different quantum oracles. The proposed algorithm, considering both types of oracles, are compared against NSGA-II, a highly cited multiobjective optimization evolutionary algorithm. Experimental evidence suggests that the quantum optimization method proposed in this work is at least as effective as NSGA-II in average, considering an equal number of executions.

1 Introduction

Many applications of the modern world rely on fast and efficient optimization algorithms. Applications in engineering, data analysis, bioinformatics and others are tackled as optimization problems where most of the time involve the simultaneous optimization of multiple conflicting objectives. The field of multiobjective (or multicriteria) optimization studies the methods for solving optimization problems with multiple objectives. For many years, researchers have been developing fast algorithms for real world applications with two or more objective functions with a lot of success [1].

In this work we present an empirical study of a proposal for a quantum algorithm for multiobjective optimization. The formal study of constructing quantum algo-

M. Villagra is supported by CONACYT research grant PINV15-208.

B. Barán · M. Villagra (✉)
Universidad Nacional de Asunción, NIDTEC,
Campus Universitario, San Lorenzo C.P.2619, Paraguay
e-mail: mvillagra@pol.una.py

B. Barán
e-mail: bbaran@pol.una.py

rithms for multiobjective optimization problems is very recent [2], and a goal of this work is to deepen our knowledge on the subject by trying to close the gap between the communities of multiobjective optimization and quantum computation.

1.1 Background

Quantum computation is a model of computation that exploits quantum phenomena to speed-up algorithms in order to find solutions of some computational problems. An important example of a quantum algorithm is the factoring algorithm of Shor [3] that can factor large composite number efficiently; any other classical algorithm for factoring seems to be intractable. Another application is in blind-search, where Grover's algorithm [4] can find a distinguished element out of a set of size N with time complexity $O(\sqrt{N})$ while any other classical algorithm requires $\Omega(N)$ time.

The field of quantum computation is currently advancing at a very fast pace mostly due to the discovery of new quantum algorithms for machine learning, big data, image processing, to name a few [5]. The key idea behind most of these applications are clever quantum algorithms for optimization problems. Quantum optimization initially became relevant with the discovery of the quantum adiabatic algorithm by Farhi et al. [6], and more recently, due to the construction of quantum annealing computers capable of handling in the order of thousand qubits [5].

Multiobjective optimization, on the other hand, is also a rich field of study with many applications in science and engineering [1]. In multiobjective optimization there are two or more conflicting objective functions, and we are interested in finding an "optimal" solution. Since the objective functions present trade-offs among them, usually there is no single optimal solution but a set of compromise solutions known as *Pareto set*. Researchers discovered several efficient algorithms that can find Pareto optimal solutions of multiobjective optimization problems and, among these algorithms, evolutionary algorithms seem to be one of the most efficient and effective approaches [1].

Even though quantum optimization is a well studied subject, and considering also the popularity of multiobjective optimization, it was only recently that a first quantum algorithm for multiobjective optimization was proposed in [2]. The algorithm of [2] shows how to map a multiobjective optimization algorithm onto the quantum adiabatic algorithm of Farhi et al. [6] and proves that a Pareto solution can be found in finite time. This reduction, however, only works in restricted cases of multiobjective problems mostly due to the requirements of the quantum adiabatic algorithm.

1.2 Contributions

In this work, we propose an extension of Grover's algorithm to multiobjetive optimization problems. Two algorithms are proposed that can query the objective functions via so-called quantum oracles. The two oracles are then implemented in an algorithm called MOGAS, from *Multiobjective Optimization Grover Adaptive Search*, which is based on the Grover adaptive search algorithm of Baritompa et al. [7].

The experimental results of this work suggest that the proposed MOGAS algorithm (considering both types of oracles) was not only an effective approach for multiobjective optimization problems, but it was also efficient when compared against NSGA-II, which is one of the most cited multiobjective algorithms [8]. In most of the studied cases, MOGAS obtained better or equal results in average for the same number of executions. It is important to note that in spite of the simple adaptive strategies used by MOGAS, the experimental results of this work present a remarkable performance over NSGA-II. Therefore, presented experimental results show the efficiency of a simple quantum algorithm with respect to NSGA-II, a representative classical algorithm for multiobjective optimization. A shorter version of this work appeared in [9].

Using Grover's algorithm for multiobjective optimization problems is in stark contrast to the adiabatic algorithm approach of [2]. In particular, Grover's algorithm is defined in the quantum circuit model of computation [4] and the adiabatic algorithm is defined on the quantum adiabatic model [6]. Both models of computation present advantages and disadvantages. For example, in the algorithm of [2], only a subset of all Pareto solutions can be found but finite-time convergence can be proved; on the other hand, with MOGAS, all Pareto solution may be found running the algorithm several times, but convergence is not guaranteed.

1.3   Outline of this Chapter

This chapter is organized as follows. Section 2 presents a review of the foundations of quantum computation and an explanation of Grover's algorithm. Section 3 briefly explains quantum optimization algorithms focusing in quantum adiabatic algorithms and the Dürr-Høyer algorithm [10]. Section 4 presents the main concepts of multiobjective optimization and briefly reviews the algorithm NSGA-II. Section 5 analyses the main proposal of this work with experimental results and discussions. Finally, Sect. 6 concludes this paper.

2   Quantum Computation

2.1   Fundamentals

A *classical* bit holds two states, 1 or 0. A quantum bit, or qubit, is defined as a vector on a Hilbert space. We denote vectors using the Dirac notation, i.e., $|\psi\rangle$ is a vector in some vector space, and $\langle\psi|$ represents its dual in the dual vector space of functionals.

Let $\mathcal{H} = span\{|0\rangle, |1\rangle\}$ be a Hilbert space equipped with a ℓ_2-norm. A qubit is a vector $|\psi\rangle \in \mathcal{H}$ over the complex field defined as

$$|\psi\rangle = \alpha |0\rangle + \beta |1\rangle,$$

where $|\alpha|^2 + |\beta|^2 = 1$.

The set $\{|0\rangle, |1\rangle\}$ is known as the *computational basis* which is an orthonormal basis for $\mathcal{H}$. These vectors can be written in matrix notation as

$$|0\rangle = \begin{bmatrix} 1 \\ 0 \end{bmatrix} \quad \text{and} \quad |1\rangle = \begin{bmatrix} 0 \\ 1 \end{bmatrix},$$

which is a canonical basis for a 2-dimensional space.

A qubit is considered as a one qubit register holding the two states $|0\rangle$ and $|1\rangle$ at the same time, something known as a *superposition state*. When a classical bit is queried, a deterministic answer (either 0 or 1) is obtained. On the contrary, when a qubit is queried, a probabilistic answer is obtained. This probabilistic answer will be 0 with probability $|\alpha|^2$ or 1 with probability $|\beta|^2$.

In order to generalize a system of n qubits, a *tensor product* is needed [11]. Let $\mathcal{H} = span\{|x\rangle : x \in \{0, 1\}^n\}$ be a Hilbert space with a ℓ_2-norm. An n qubit state is a vector $|\psi\rangle \in \mathcal{H}$ defined as

$$|\psi\rangle = \sum_{x \in \{0,1\}^n} \alpha_x |x\rangle,$$

where $\sum_x |\alpha_x|^2 = 1$, and $|x\rangle = |x_1\rangle \otimes \cdots \otimes |x_n\rangle$ with each $|x_i\rangle \in span\{|0\rangle, |1\rangle\}$ and $\otimes$ denoting a tensor product [11].

Algorithms for quantum computers are built as a sequence of quantum operations (or gates, in analogy to classical circuits) acting on qubit registers. These quantum operations are essentially unitary operations defined over a Hilbert space.

An important quantum gate is the Hadamard operation denoted with $\mathbf{H}$. It is basically a rotation operation acting in the following way

$$\mathbf{H} |i\rangle = \frac{1}{\sqrt{2}} (|0\rangle + (-1)^i |1\rangle),$$

for $i \in \{0, 1\}$. This operation is defined in matrix form as

$$\mathbf{H} = \frac{1}{\sqrt{2}} \begin{bmatrix} 1 & 1 \\ 1 & -1 \end{bmatrix}.$$

In an n-qubit system, the number of rows and columns in the matrix representation of operators grow exponentially.

Evolution in a closed quantum system is unitary, that is, given a quantum register $|\psi\rangle$ and an unitary operation U, the state of the system in the next time-step is $U |\psi\rangle$.

At the end of a quantum computation, to read-out the result, we need to be able to measure the state of the system. To explain how a measurement works, first we start by defining an *observable*. In simple terms, an observable is a dynamic variable we want to measure, e.g., velocity, energy, spin, etc. Normally, in quantum computation we want to know if a qubit is in state $|0\rangle$ or $|1\rangle$. Formally, an observable is a Hermitian operator that acts on the Hilbert space of the system whose eigenvalues and eigenvectors correspond to the values and states of the dynamic variable.

To be able to measure an observable we need to apply a *measurement* operator. A measurement is a set of linear operators $\{M_m\}$ that acts on the Hilbert space of the system being observed. The index m refers to the outcome of the measurement. Say that the system is in state $|\psi\rangle$. When we measure $|\psi\rangle$, the probability that m occurs is

$$P(m) = \langle\psi| M_m^\dagger M_m |\psi\rangle,$$

and the state of the system after measurement is

$$\frac{M_m |\psi\rangle}{\sqrt{P(m)}}.$$

The simplest type of measurements are *projective measurements*. Given an observable O, its spectral decomposition is

$$O = \sum_m mP_m,$$

where P_m is a projection onto the subspace with eigenvalue m. Thus, the probability of getting m is

$$P(m) = \langle\psi| P_m |\psi\rangle,$$

and the state of the system after measurement is

$$\frac{P_m |\psi\rangle}{\sqrt{P(m)}}.$$

2.2 Grover's Search Algorithm

The first quantum algorithm for unstructured search was discovered by Grover [4]. Given N objects, with no information about how they are positioned in the search space, Grover's algorithm finds an element from a set of marked objects with high probability. Grovers algorithm has a running-time complexity of $O(\sqrt{N})$ for a set of N objects. This is a quadratic speed-up with respect to classical algorithms (classically for a randomized algorithm $\Omega(N)$ steps are required). This bound is tight as showed by Bennett et al. [12].

The field of quantum search algorithms is a very popular area. Several problems in the query model of computation were developed for graphs, matrices, groups, etc. (see *The Quantum Algorithms Zoo*[1]).

In this section, Grover's algorithm is briefly explained, given that it is an integral part of the proposed algorithm. For details refer to the book by Nielsen and Chuang [11].

In classical computation, finding a specific element out from a set of N elements requires N tries. Grover's search algorithm, however, can find a specific element out from a finite set of N elements with complexity $O(\sqrt{N})$. This is possible because of quantum interference, which the algorithm exploits via a quantum operator $\mathbf{G}$ known as the *Grover operator*. The Grover operator is constructed from an oracle operator $\mathbf{O_G}$ and a phase operator $\mathbf{W}$. The number of iterations r sufficient to find a desired item out of N alternatives with high probability is $r = \left\lfloor \frac{\pi}{4}\sqrt{N} \right\rfloor$ [4].

The input to Grover's algorithm is a set of n qubits $|0\rangle^{\otimes n}$, where $2^n = N$, and an ancilla qubit $|1\rangle$. The first input $|0\rangle^{\otimes n}$ is transformed to a superposition state using an n-fold Hadamard transformation $\mathbf{H}^{\otimes n}$,

$$|\zeta\rangle = \mathbf{H}^{\otimes n}|0\rangle^{\otimes n} = \frac{1}{\sqrt{N}} \sum_{x \in \{1,0\}^n} |x\rangle, \tag{1}$$

where $\mathbf{H}^{\otimes n} = \mathbf{H} \otimes \cdots \otimes \mathbf{H}$, n times. Analogously, $|0\rangle^{\otimes n} = |0\rangle \otimes \cdots \otimes |0\rangle$, n times.
The second register is transformed using a Hadamard gate according to

$$\mathbf{H}|1\rangle = |-\rangle = \frac{|0\rangle - |1\rangle}{\sqrt{2}}. \tag{2}$$

Grover's algorithm is based on the ability of an oracle to "mark" a desired solution, which is represented by one of the basis states. Given a superposition state, the marking process of an oracle is a change of the sign of the coefficient in the basis state which corresponds to a desired solution; such a marking process will only be possible if some interaction exists between the oracle operator and the ancilla register. After the marking process, the phase operator makes an increase of the absolute value of the amplitude associated to the solution state while decreasing amplitudes associated to the other non-solution states via quantum interference. If properly done, this will happen at each iteration, and with the help of interference, it is possible to observe/measure the desired solution state with high probability [11].

Grover's algorithm can be generalized to find a set of M distinguished elements [12]. The number of oracle queries thus becomes $r = O(\sqrt{(N/M)})$. Hence, the optimal number of queries to the quantum oracle depends on the number of distinguished elements.

---

[1] http://math.nist.gov/quantum/zoo/

3 Quantum Optimization

3.1 Adiabatic Algorithms

Adiabatic quantum computation (AQC) appeared as an alternative to the quantum gate model of computation with the goal of solving optimization problems. AQC takes its name from an early theorem by Born and Fock [13] known as the quantum adiabatic theorem which states that given a slowly changing quantum system represented by a Hamiltonian that depends on time, if the system starts in its ground-state, then after a sufficiently large time, even though the Hamiltonian changed, the system remains in its time-dependent ground-state.

AQC was discovered by Farhi et al. [6], and the authors showed how to exploit the adiabatic evolution of a quantum system to find optimal solutions of an optimization problem. AQC also found applications in machine learning, combinatorics, protein folding, to name a few [5].

Recently, Barán and Villagra [2] showed how to solve a multiobjective optimization problem using AQC. The approach of [2], however, works only for a restricted class of multiobjective optimization problems and cannot find all potential solutions.

3.2 Dürr and Høyer's Algorithm

Grover's algorithm is generally used as a search method to find a set of desired solutions from a set of possible solutions. However, Dürr and Høyer presented an algorithm based on Grover's method for optimization [10]. Their algorithm finds an element of minimum value inside an array of N elements using in the order of $O(\sqrt{N})$ queries to the oracle.

Baritompa et al. [7] presented an application of Grover's algorithm for global optimization, which they call *Grover Adaptative Search (GAS)*. Basically, GAS is based on Grover's search with an "adaptive" oracle operator in a minimization context of the objective function. The oracle operator marks all the solutions from a set below a certain threshold value y given by

$$g_y(x) = \begin{cases} 1, & \text{if } f(x) < y \\ 0, & \text{if } f(x) \geq y \end{cases}, \tag{3}$$

where x is a possible solution in the decision space and $f(x)$ is the value of the objective function (in this case, the value of the objective function of a current known solution y). The oracle marks a solution x if and only if the boolean function $g_y(x) = 1$.

Algorithm 1 Dürr and Høyer's Algorithm

1: Randomly choose x from the decision space.
2: Set $x_1 \leftarrow x$.
3: Set $y_1 \leftarrow f(x_1)$.
4: Set $m \leftarrow 1$.
5: Choose a value for the parameter λ (8/7 is suggested in [7]).
6: For $k = 1, 2, \ldots$ until termination condition is met, do:

    (a) Choose a random rotation count r_k uniformly from $\{0, \ldots, \lceil m - 1 \rceil\}$.
    (b) Apply Grover search with r_k iterations on $f(x)$ with threshold y_k, and denote the outputs by x and y.
    (c) If $y < y_k$, set $x_{k+1} \leftarrow x$, $y_{k+1} \leftarrow y$ and $m \leftarrow 1$; otherwise, set $x_{k+1} \leftarrow x_k$, $y_{k+1} \leftarrow y_k$ and $m \leftarrow \min\{\lambda m, \sqrt{N}\}$.

The algorithm requires two extra parameters, a currently known solution x_k and an iteration count r_k. The iteration count r_k is a value computed from the number of solutions that are better than the currently known solution. Initially, the algorithm randomly chooses a feasible solution x from the decision space which becomes the known solution; the number of solutions that are better than y, however, is unknown and an iteration count is required to perform the search. This is due to the black box nature of the oracle [7].

When the algorithm finds a better solution, it becomes the new known solution. This solution is then used as a new threshold for the next iteration of GAS and the sequence of iteration counts must be computed again. In this way, GAS can find improved solutions in an adaptive searching framework.

Dürr and Høyer introduced a strategy for the selection of the iteration count based on a random selection of a number from a set of integer numbers. This set starts with $\{0\}$ as the only element. When the search is unsuccessful in finding a better solution, the algorithm adds more elements to a maximum of $\{0, \ldots, \lceil m - 1 \rceil\}$ at each search step, until a solution better than the current known solution is found. In this way, the set incorporates more integer numbers as elements. Thus, the probability of selecting the right iteration count for a successful search increases.

The value of m is updated at each step by $\min\{\lambda^i m, \sqrt{N}\}$, where λ is given as a parameter, i represents the count of the previous unsuccessful search steps and $N = 2^n$ is the number of total elements from the decision space based on the number of qubits n. Therefore, m is not allowed to exceed $\sqrt{N}$, which is the optimal iteration number to find a specific element from a set of N elements.

The pseudocode of Dürr and Høyer's algorithm based on the GAS algorithm is presented in Algorithm 1. This corresponds to an interpretation that has been described by Baritompa et al. [7], where the parameter k represents the search process count.

4 Multiobjective Optimization

4.1 Fundamentals

A general multiobjective optimization Problem (MOP) is defined with a set of n decision variables, k objective functions, and m restrictions, that is,

$$\text{Optimize } y = f(x) = (f_1(x), \ldots, f_k(x))$$
$$\text{subject to } \gamma(x) = (\gamma_1(x), \ldots, \gamma_m(x)) \geq 0,$$

where $x \in \mathbb{R}^n$ is called the decision vector, and y is the objective vector. We also let X be the decision space and Y is the objective space. Depending on the problem, optimize could mean minimize or maximize. The set of restrictions $\gamma(x) \geq 0$ determines the set of feasible solutions $X_f \subseteq X$ and the set of objective vectors $Y_f \subseteq Y$. A MOP is the problem of finding a feasible solution x that optimizes y. In general, there is no unique best solution but a set of solutions, none of which can be considered better than the others when all objectives are considered at the same time. This comes from the fact that there can be conflicting objectives. Thus, a new concept of optimality should be established for MOPs.

Given two decision vectors $u, v \in X$ we define

$$f(u) = f(v) \text{ iff } f_i(u) = f_i(v) \text{ for all } i = 1, \ldots, k,$$
$$f(u) \leq f(v) \text{ iff } f_i(u) \leq f_i(v) \text{ for all } i = 1, \ldots, k, \text{ and}$$
$$f(u) < f(v) \text{ iff } f(u) \leq f(v) \text{ and } f(u) \neq f(v).$$

Then, in a minimization context, we say that u dominates v, denoted $u \prec v$, if $f(u) < f(v)$. We also say that u and v are non-comparable, denoted $u \smile v$, when neither u dominates v nor v dominates u.

When x is non-dominated with respect to the entire set X_f, we say that x is a *Pareto-optimal solution*. Thus, we define the Pareto optimal set X_p as the set of non-dominated solutions with respect to the set of feasible solutions X_f. The corresponding set of objective vectors $Y_p = f(X_p)$ is known as Pareto front.

4.2 NSGA-II

Evolutionary algorithms imitate the behavior of biological evolution in order to search for good quality solutions of a given problem. In order to find optimal solutions, evolutionary algorithms define so-called *individuals* as posibles solutions to an optimization problem and *populations* as sets of individuals. Each individual is assigned a *fitness* value, which measures the quality of the individual as a solution. Furthermore, evolutionary algorithms make use of a set of operators called *evolutionary operators*. These operators are the selection, crossover and mutation operators.

Evolutionary operators are commonly used with random elements to generate new individuals which make up a population at each new generation. Poor individuals are usually discarded by this process when creating a new generation. Thus, evolutionary algorithms evolve toward good quality solutions.

NSGA-II, from *Non-dominated Sorting Genetic Algorithm—version two*, is among the most cited evolutionary algorithms [1, 8] and works in the following way. Given an initial population P_0 of size N, first P_0 is classified by a procedure called *fast non-dominated sorting*. This procedure classifies solutions according to the number of solutions which dominate them. After the classification process, each individual is assigned a fitness value according to its non-dominance level. Then, good solutions (based on their fitness value) are selected from P_0 and the evolutionary operators of crossover and mutation are applied to these selected solutions in order to produce the next generation called Q_0. Once the populations P_0 and Q_0 are available, the algorithm enters its main loop shown in Algorithm 2. Each iteration t of the main loop takes two populations as input, one parent population P_t and one child population Q_t. With these two populations, a new population R_{t+1} of size at most $2N$ is created from the union of P_t and Q_t. The population R_{t+1} allows a comparison between the best solutions from P_t and Q_t. The new population R_{t+1} is then classified by non-dominance and then the new population P_{t+1} is created with the best solutions from P_t and Q_t. NSGA-II makes use of a *crowding distance* to determine how isolated a solution is from other solutions. The algorithm thus gives preference to more isolated individuals when comparing two solutions which do not dominate each other to generate P_{t+1} using a *crowding operator* $\prec_n$. If u and v are two solutions, we say that $u \prec_n v$ if u has a better fitness than v, and in case of a tie, NSGA-II selects the solution that is most isolated in lexicographic order. Finally, a population Q_{t+1} is created by selecting good solutions from P_{t+1} and applying the operators of crossover and mutation to create new children until Q_{t+1} is of size N. The populations P_{t+1} and Q_{t+1} are then used as input for the next iteration of the main loop.

A stopping criterion is used to halt the main loop, for example, the number of generations. At the end of an execution, an approximated Pareto front is obtained made of solutions which are not dominated by any other solution from the last generation. More details about NSGA-II can be found in [8].

5 Multiobjective Grover Adaptive Search

5.1 The Algorithm

In this work, a new adaptative search algorithm based on the heuristic of Dürr and Høyer is proposed named *Multiobjetive Grover Adaptive Search (MOGAS)*. MOGAS may use two different oracle operators based on the Pareto dominance relation. The first oracle h_1 marks all non-dominanted solutions with respect to a known

---

Algorithm 2 Main loop of NSGA-II.

1: $R_t \leftarrow P_t \cup Q_t$.
2: $\mathscr{F} \leftarrow$ fast-non-dominated-sort(R_t).
3: Let $P_{t+1} \leftarrow \emptyset$ and $i \leftarrow 1$.
4: Repeat while $|P_{t+1}| + |\mathscr{F}| \leq N$

    (a) crowding-distance-assignment(F).
    (b) $P_{t+1} \leftarrow P_{t+1} \cup \mathscr{F}$.
    (c) Increment i in one.

5: sort($F, \prec_n$).
6: Let P_{t+1} be made of the individuals from P_{t+1} plus $|\mathscr{F}| - |P_{t+1}|$ individuals from $\mathscr{F}$.
7: $Q_{t+1} \leftarrow$ make-new-population(P_{t+1}).

---

current solution. The second oracle h_2 marks all non-dominated and non-comparable solutions. These oracles are based on the boolean functions

$$h_1(x) = \begin{cases} 1 & \text{if } f(x) \prec y, \\ 0 & \text{otherwise,} \end{cases} \tag{4}$$

$$h_2(x) = \begin{cases} 1 & \text{if } f(x) \prec y \vee f(x) \sim y, \\ 0 & \text{otherwise,} \end{cases} \tag{5}$$

where x is a feasible solution of the decision space, $f(x)$ is a vector where each element represents the value of the objective function with respect to solution x, and y is a vector where each element is the value of each objective function for the current known solution.

The first oracle marks a non-dominated solution if and only if the boolean function $h_1(x) = 1$. In a similar way, the second oracle marks a non-dominated or non-comparable solution if and only if the boolean function $h_2(x) = 1$.

The pseudocode of MOGAS algorithm is presented below in Algorithm 3, where the parameter k represents the search process count.

The operation of MOGAS is based on the oracle operator. Then, using any of the presented oracles, h_1 or h_2, MOGAS can find a non-dominated solution with respect to a known solution. In this way, the algorithm can reach the Pareto-optimal set by finding new non-dominated solutions at each iteration. Therefore, with the proposed search process it is possible to incorporate a new element into the Pareto-optimal set or replace old elements from it each time a non-dominated solution is found.

5.2 Experimental Results

Currently, a general purpose quantum computer has not been yet implemented. Nevertheless, the basic ideas of quantum algorithms can be fully explored using linear algebra, and therefore, computational performances of quantum algorithms are possible by executing linear algebra operations [14].

Algorithm 3 MOGAS Algorithm

1: Randomly choose x from the decision space.
2: Set $S \leftarrow \{x_1 \leftarrow x\}$
3: Set $y_1 \leftarrow f(x_1)$.
4: Set $m \leftarrow 1$.
5: Choose a value for the parameter λ (8/7 is suggested).
6: For $k = 1, 2, \ldots$ until termination condition is met, do:

   (a) Choose a random rotation count r_k uniformly from $\{0, \ldots, \lceil m - 1 \rceil\}$.
   (b) Perform a Grover search of r_k iterations on $f(x)$ with threshold t_k, and denote the outputs
       by x and y.
   (c) If $y \not\prec y_k$ set $x_{k+1} \leftarrow x_k$, $t_{k+1} \leftarrow t_k$ and $m \leftarrow \min\{\lambda m, \sqrt{N}\}$. Otherwise, set $m \leftarrow 1$,
       $x_{k+1} \leftarrow x$, $t_{k+1} \leftarrow t$ and with respect to all elements of the set S, where $j = 1, \ldots, |S|$, do:
       If $\exists\, x_j \in S$ such that $f(x) \prec f(x_j)$, then set $S \leftarrow S - \{x_j\}$ and finally set $S \leftarrow S \cup \{x\}$.

7: Set $PF \leftarrow \{f(x_j) : j = 1, \ldots, |S|\}, \forall\, x_j \in S$.

To verify the effectiveness of MOGAS, we compared it against the already presented NSGA-II algorithm. The tests were made considering some biobjective problems based on the well known ZDT test suite [15] and randomly generated (RG) instances.

The randomly generated problems (RG) consist of a random selection of numbers from a set of integer numbers between 1 and 1000 for each of the two objective functions. Then, three different suites of this type of random instances were established for testing. With respect to the ZDT test suite, the ZDT1, ZDT3 and ZDT4 were selected considering two objective functions. For each of these functions, a total of twenty decision variables were used and to each of these decision variables a random real number from the interval [0, 1] was assigned.

The decision space for each instance is a set of $1024 = 2^{10}$ points. The amount of points is based on the number of qubits ($n = 10$) selected for the proposed MOGAS algorithm. Since the problem has two objective functions that must be minimized, the vector dimension for $f(x)$ and y is $p = 2$. Table 1 presents the main characteristics of the test suites considered.

The testing procedure was based on ten executions of both algorithms, that is, MOGAS (considering the two different types of oracles) and NSGA-II, over all test suites. At each execution, the termination criteria was to complete two hundred generations (with a population size equal to fifty) for NSGA-II and a total of four hundred algorithm consultations for MOGAS, where a algorithm consultation corresponds to an execution of Grover's search with regard to r_k iterations on $f(x)$ considering a threshold y_k, and denoting the outputs by x and y respectively.

The hypervolume was used as a metric for comparison of the results, also because it is the most used comparison metric in multiobjective optimization [16]. The hypervolume is an indicator used in the multiobjective optimization of evolutionary algorithms to evaluate the performance of the search proposed by Zitzler and Thiele [17]. It is based on a function that maps the set of Pareto-optimal solutions to a scalar with

Table 1 Test suites for the experiments

Function	m	x_i, x_j $i, j = 1, \ldots, 2^{10}$	f_1	f_2
RG$_{1,2,3}$	–	$x_i \in [1, 10^3]\, x_j \in [1, 10^3]\, x_i, x_j \in \mathbb{N}$	x_i	x_j
ZDT1	20	$x_i \in [0, 1]$	x_i	$g_1(x_i)\lfloor 1 - \sqrt{x_i/g_1(x_i)}\rfloor,$ $g_1(x_i) = 1 + 9\frac{(\sum_{k=2}^m x_{i_k})}{(m-1)}$
ZDT3	20	$x_i \in [0, 1]$	x_i	$g_3(x_i)\lfloor 1 - \sqrt{x_i/g_3(x_i)}$ $-\frac{x_i}{g_3(x_i)}\sin(10\pi x_i)\rfloor,$ $g_3(x_i) = 1 + 9\frac{(\sum_{k=2}^m x_{i_k})}{(m-1)}$
ZDT4	20	$x_i \in [0, 1]$	x_i	$g_4(x_i)\lfloor 1 - \sqrt{x_i/g_4(x_i)}\rfloor,$ $g_4(x_i) = 1 + 10(m-1)+$ $\sum_{k=2}^m (x_{i_k}^2 - 10\cos(4\pi x_{i_k}))$

respect to a reference point. Given that it is used in a minimization context, larger hypervolume value represents a better solution.

In Tables 2, 3 and 4, the experimental results obtained from the testing procedure are presented considering the hypervolume. The tables are composed of six columns that correspond to each test suite and a column for the order of execution. In these six columns, the result of the hypervolume metric in percentage for each execution is given. In this way, each row summarizes the experimental results for every test suite with respect to a specific execution order denoted in the left column. Also, in the last row, an average of these ten executions for all test suites is presented.

Table 2 MOGAS, after 400 consultations using h_1

# Runs	Test sets					
	RG$_1$	RG$_2$	RG$_3$	ZDT1	ZDT3	ZDT4
1	98.4	98.4	98.8	51.4	57	61.2
2	99	98.4	99.1	52.8	57.3	60.2
3	99	98.6	98.6	52.7	58.1	60.5
4	99	98.7	99.1	52.1	58.2	60.7
5	98.9	98.9	99.1	52.7	58.2	60.5
6	99.1	98.5	98.7	52.4	58.3	60.7
7	98.9	98.5	99	52.9	57.1	61.3
8	99.1	98.7	99.1	53.1	56.5	58.8
9	98.9	98.7	99.1	52.3	58.2	59.7
10	98.9	98.7	98.4	53.2	57.7	58.9
Average	99	99	99	53	58	60

Table 3 MOGAS, after 400 consultations using h_2

# Runs	Test sets					
	RG_1	RG_2	RG_3	ZDT1	ZDT3	ZDT4
1	98	98.7	99	51	55.4	57.4
2	96.6	98.4	99.1	49	56.4	59.9
3	98.7	98.7	99.1	50.7	53.2	60.4
4	97.3	98.2	99.2	50.4	53.8	61.1
5	98.7	98.7	98.7	50.9	55.4	55.1
6	99.2	98.9	96.7	50.7	54.2	58
7	98.4	98.7	99	49.7	52.7	59.8
8	98.8	98.2	98.6	49	52	59.1
9	98.1	98.8	97.6	47.7	52.3	61.3
10	97	98.6	99.2	49.1	53.2	58.9
Average	98	99	99	50	54	59

Table 4 NSGA-II, after 200 generations and a population size equal to 50

# Runs	Test sets					
	RG_1	RG_2	RG_3	ZDT1	ZDT3	ZDT4
1	98.1	97.3	98.4	52.1	55.7	60.2
2	99	96.8	97.7	51.2	56.4	60.1
3	97.8	98.6	97.5	51.1	56.9	60.1
4	97.1	98.1	99.1	51.9	55.6	59.6
5	97.5	97.1	98.3	51.9	58.4	60.4
6	98.2	96.6	98.5	52.7	56.6	59.7
7	97.9	98.2	98.8	53.2	57.6	60.7
8	97.6	97.7	98.8	51.5	57.2	60.6
9	97.8	96.1	98.8	51.9	55.9	60
10	98.7	97.3	98.5	52.8	58	59.6
Average	98	97	98	52	57	60

Tables 2 and 3 correspond to results obtained for MOGAS using h_1 and h_2 respectively. Table 4 corresponds to results obtained using NSGA-II with a population size equal to fifty.

From the experimental results, MOGAS presents similar results compared to NSGA-II with a population size of fifty with respect to RG problems; in most cases, however, MOGAS delivers better or at least equal results. However, considering the structured ZDT test suites and compared to NSGA-II results, only MOGAS based on the boolean function h_1 as oracle presents equal or better results, whereas MOGAS based using h_2 as oracle presents nearly equal results but not equal or better results. Furthermore, considering the algorithm consultations of MOGAS as a single evaluation of the objective function, the results present an important fact to note: MOGAS used only four hundred evaluations of the objective function vector $f(x)$, whereas NSGA-II (with a population size of fifty) used 10,000 evaluations (population size times number of generations = 50 * 200) of the same objective vector to deliver similar results.

Tables 5, 6 and 7 summarize the average results of both MOGAS algorithms and NSGA-II, considering objective function evaluations. These tables are composed of six columns that correspond to each test suite and a column for the number of evaluations. In these six columns, the average results of the hypervolume metric in percentage corresponding to ten executions are presented. In this way, each row summarizes the average results for every test suite with respect to a specific number of evaluations given in the left column.

Table 5 Average Results of the testing procedure—MOGAS. From 100 to 400 evaluations and the oracle based on the boolean function h_1

# Eval.	Test sets					
	RG_1	RG_2	RG_3	ZDT1	ZDT3	ZDT4
100	97.7	97.4	98.2	49.2	54.8	57.9
200	98.5	98.2	98.6	51.4	56.8	58.9
300	98.9	98.5	98.8	52.3	57.5	59.7
400	98.9	98.6	98.9	52.5	57.7	60.2

Table 6 Average results of the testing procedure—MOGAS. From 100 to 400 evaluations and the oracle based on the boolean function h_2

# Eval.	Test sets					
	RG_1	RG_2	RG_3	ZDT1	ZDT3	ZDT4
100	94.2	95.1	92.9	44.7	47	55.1
200	97	97.7	97	47.6	50.2	57.4
300	97.9	97.7	98.2	48.7	52.7	58.4
400	98.1	98.6	98.6	49.8	53.9	59.1

Table 7 Average results of the testing procedure—NSGA-II. From 100 to 10,000 evaluations corresponding to a population size equal to 50

# Eval.	Test sets					
	RG_1	RG_2	RG_3	ZDT1	ZDT3	ZDT4
100	94.6	94.5	95.6	47.4	51	54.9
200	95.9	95.1	96.2	49	51.9	56.8
300	96.5	95.3	96.5	49.4	53	57.3
400	96.5	95.9	96.8	49.9	53.4	57.7
4000	97.7	97.2	98.1	51.5	56.5	60
10000	98	97.4	98.4	52	56.8	60.1

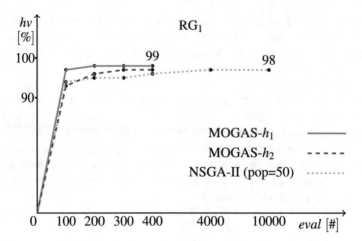

Fig. 1 Graphs of the hypervolume metric in percentage (*hv*) versus the number of evaluations of the objective function vector (*eval*) made by each algorithm (MOGAS and NSGA-II) with respect to the RG$_1$ suite test

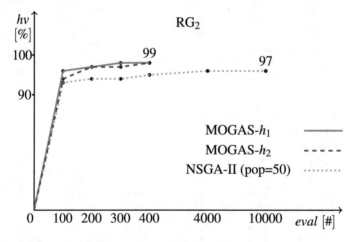

Fig. 2 Graphs of the hypervolume metric in percentage (*hv*) versus the number of evaluations of the objective function vector (*eval*) made by each algorithm (MOGAS and NSGA-II) with respect to the RG$_2$ suite test

The experimental results are presented in Figs. 1, 2, 3, 4, 5 and 6 as performance using the hypervolume metric (in percentage) versus the number of evaluations of the objective function vector.

Considering the average number of iterations of the Grover operator needed for MOGAS using both oracles, the presented experimental results reveal that MOGAS using h_2 as oracle, in most cases, uses less iterations than MOGAS using h_1 as oracle.

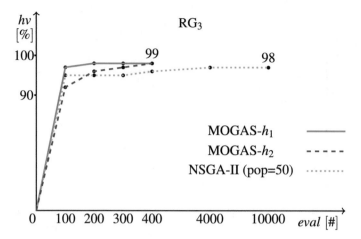

Fig. 3 Graphs of the hypervolume metric in percentage (*hv*) versus the number of evaluations of the objective function vector (*eval*) made by each algorithm (MOGAS and NSGA-II) with respect to the RG$_3$ suite test

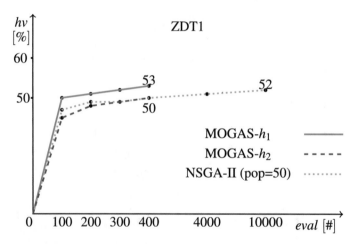

Fig. 4 Graphs of the hypervolume metric in percentage (*hv*) versus the number of evaluations of the objective function vector (*eval*) made by each algorithm (MOGAS and NSGA-II) with respect to the ZDT1 suite test

Certainly, the oracle based on h_2 marks more solutions from the decision space. Therefore, the probability to change the threshold at every consultation of Grover's algorithm increases. This way, the parameter m is set to 1 more often and the chosen iteration number corresponds to a lower number. Thus, the total number of iterations for MOGAS using h_2 is smaller than the case when using the oracle based on h_1.

Tables 8, 9, 10, 11, 12 and 13 summarize the average results of the number of iterations used by MOGAS, considering the number of times the Grover operator is invoked. These tables have two columns that correspond to each different type of

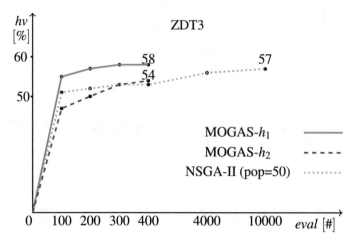

Fig. 5 Graphs of the hypervolume metric in percentage (*hv*) versus the number of evaluations of the objective function vector (*eval*) made by each algorithm (MOGAS and NSGA-II) with respect to the ZDT3 suite test

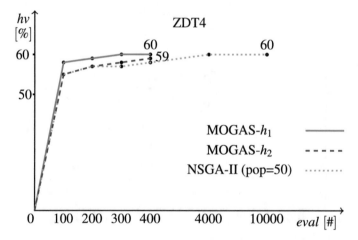

Fig. 6 Graphs of the hypervolume metric in percentage (*hv*) versus the number of evaluations of the objective function vector (*eval*) made by each algorithm (MOGAS and NSGA-II) with respect to the ZDT4 suite test

oracle and a column for the number of evaluations. In these two columns, the average results of the number of iterations corresponding to ten executions are presented. In this way, each row summarizes the average result for both oracles with respect to a specific number of evaluations presented in the left column.

Table 8 Average iteration numbers used for RG1. From 100 to 400 evaluations

# Eval.	Oracle	
	MOGAS-h_1	MOGAS-h_2
100	815	352
200	2162	1299
300	3588	2277
400	5149	3474

Table 9 Average iteration numbers used for RG2. From 100 to 400 evaluations

# Eval.	Oracle	
	MOGAS-h_1	MOGAS-h_2
100	748	343
200	2078	991
300	3483	2373
400	4975	3485

Table 10 Average iteration numbers used for RG3. From 100 to 400 evaluations

# Eval.	Oracle	
	MOGAS-h_1	MOGAS-h_2
100	838	349
200	1952	888
300	3344	1947
400	4848	3274

Table 11 Average iteration numbers used for ZDT1. From 100 to 400 evaluations

# Eval.	Oracle	
	MOGAS-h_1	MOGAS-h_2
100	219	280
200	602	801
300	1182	1517
400	2094	2385

Table 12 Average iteration numbers used for ZDT3. From 100 to 400 evaluations

# Eval.	Oracle	
	MOGAS-h_1	MOGAS-h_2
100	255	259
200	863	668
300	1635	1319
400	2571	2247

Table 13 Average iteration numbers used for ZDT4. From 100 to 400 evaluations

# Eval.	Oracle	
	MOGAS-h_1	MOGAS-h_2
100	407	410
200	1260	1255
300	2676	2273
400	3858	3474

6 Concluding Remarks

In this work we proposed two different types of oracles used in a quantum algorithm for multiobjective optimization problems. The presented multiobjective quantum algorithm, called MOGAS, is a natural extension of previous quantum algorithms for single-objective optimization based on Grover's search algorithm. The experimental results of this work suggests that MOGAS (considering both types of oracles) was not only an effective approach for multiobjective optimization problems, but it was also efficient as was observed when MOGAS was compared against NSGA-II, which is one of the most cited multiobjective optimization algorithms [8]. In most of the studied cases, MOGAS obtained better or similar results in average after comparing it against NSGA-II for the same number of executions, primarily with respect to the oracle based on the boolean function h_1; in regard of h_2, the results presented in this work are almost equal compared to NSGA-II.

In spite of the simple adaptive strategy used by MOGAS, the experimental results of this work presented a remarkable performance over NSGA-II when considering executions with the same evaluation number of the objective functions. Therefore, the presented experimental results show the efficiency of a simple quantum algorithm with respect to a classical more elaborated algorithm.

Another interesting fact to note is the difference between the number of iterations used by MOGAS. The oracle based on the boolean function h_2, in most cases, utilized a smaller number of iterations than the one using h_1. Hence, h_2 is more efficient than h_1, which represents a saving in the number of queries to the quantum oracle.

For future research, it is interesting to study other different definitions of oracles for multiobjective problems. It is also very important to lay some theoretical foundations that can show the convergence of MOGAS to the set of Pareto-optimal solutions.

References

1. von Lücken, C., Barán, B., Brizuela, C.: A survey on multi-objective evolutionary algorithms for many-objective problems. Comput. Optim. Appl. **58**(3), 707–756 (2014)
2. Barán, B., Villagra, M.: Multiobjective optimization in a quantum adiabatic computer. Electron. Notes Theor. Comput. Sci. **329**, 27–38 (2016)
3. Shor, P.W.: Algorithms for quantum computation: discrete logarithms and factoring. In: Proceedings of 35th Annual Symposium on Foundations of Computer Science (FOCS), pp. 124–134 (1994)
4. Grover, L.K.: A fast quantum mechanical algorithm for database search. In: Proceedings of the 28th Annual ACM Symposium on Theory of computing (STOC), pp. 212–219 (1996)
5. McGeoch, C.C.: Adiabatic Quantum Computation and Quantum Annealing. Synthesis Lectures on Quantum Computing. Morgan & Claypool (2014)
6. Farhi, E., Goldstone, J., Gutman, S., Sipser, M.: Quantum Computation by Adiabatic Evolution (2000). arXiv:quant-ph/0001106
7. Baritompa, W.P., Bulger, D.W., Wood, G.R.: Grover's quantum algorithm applied to global optimization. SIAM J. Optim. **15**(4), 1170–1184 (2005)
8. Deb, K., Pratap, A., Agarwal, S., Meyarivan, T.: A fast and elitist multiobjective genetic algorithm: NSGA-II. IEEE Trans. Evol. Comput. **6**(2), 182–197 (2002)
9. Fogel, G., Barán, B., Villagra, M.: Comparison of two types of quantum oracles based on Grovers adaptative search algorithm for multiobjective optimization problems. In: Proceedings of the 10th International Workshop on Computational Optimization, Federated Conference in Computer Science and Information Systems (FedCSIS), ACSIS 11, pp. 421–428, Prague, Czech Republic, 3–6 Sept 2017
10. Dürr, C., Høyer, P.: A quantum algorithm for finding the minimum (1996). arXiv:quant-ph/9607014
11. Nielsen, M.A., Chuang, I.L.: Quantum Computation and Quantum Information. Cambridge university press (2010)
12. Bennett, C.H., Bernstein, E., Brassard, G., Vazirani, U.: Strengths and weaknesses of quantum computing. SIAM J Comput **26**(5), 1510–1523 (1997)
13. Born, M., Vladimir Fock, F.: Beweis des adiabatensatzes. Z. für Phys. **51**(3–4), 165–180 (1926)
14. Lipton, R.J., Regan, K.W.: Quantum Algorithms via Linear Algebra. MIT Press (2014)
15. Chase, N., Rademacher, M., Goodman, E., Averill, R., Sidhu, R.: A benchmark study of multi-objective optimization methods. BMK-3021, Rev 6 (2009)
16. Riquelme, N., Baran, B., von Lücken, C.: Performance metrics in multi-objective optimization. In: Proceedings of the 41st Latin American Computing Conference (CLEI) (2015)
17. Zitzler, E., Thiele, L.: Multiobjective evolutionary algorithms: a comparative case study and the strength Pareto approach. IEEE Trans. Evol. Comput. **3**(4), 257–271 (1999)

Discovering Knowledge
from Predominantly Repetitive Data
by InterCriteria Analysis

Olympia Roeva, Nikolay Ikonomov and Peter Vassilev

Abstract In this paper, InterCriteria analysis (ICrA) approach for finding existing or unknown correlations between multiple objects against multiple criteria is considered. Five different algorithms for InterCriteria relations calculation, namely μ-biased, Balanced, ν-biased, Unbiased and Weighted, are compared using a new cross-platform software for ICrA approach – ICrAData. The comparison have been done based on numerical data from series of model parameter identification procedures. Real experimental data from an *E. coli* fed-batch fermentation process are used. In order to estimate the model parameters (μ_{max}, k_S and $Y_{S/X}$) fourteen differently tuned Genetic algorithms are applied. ICrA is executed to evaluate the relation between the model parameters, objective function value and computation time. Some useful conclusions with respect to the selection of the appropriate ICrA algorithm for a given data are established. The considered example illustrates the applicability of the ICrA algorithms and demonstrates the correctness of the ICrA approach.

Keywords InterCriteria Analysis · Weighted algorithm · Genetic algorithms
E. coli · Fermentation process

O. Roeva (✉) · P. Vassilev
Institute of Biophysics and Biomedical Engineering,
Bulgarian Academy of Sciences, Acad. G. Bonchev Str., Bl. 105, 1113 Sofia, Bulgaria
e-mail: olympia@biomed.bas.bg

P. Vassilev
e-mail: peter.vassilev@gmail.com

N. Ikonomov
Department of Analysis, Geometry and Topology, Institute of Mathematics
and Informatics, Bulgarian Academy of Sciences,
8 Acad. G. Bonchev Str., 1113 Sofia, Bulgaria
e-mail: nikonomov@math.bas.bg

© Springer Nature Switzerland AG 2019
S. Fidanova (ed.), *Recent Advances in Computational Optimization*,
Studies in Computational Intelligence 795,
https://doi.org/10.1007/978-3-319-99648-6_12

1 Introduction

InterCriteria Analysis (ICrA) is an approach aiming to go beyond the nature of the criteria involved in a process of evaluation of multiple objects against multiple criteria, and, thus to discover some dependencies between the ICrA criteria themselves [1]. Initially, ICrA has been applied for the purposes of temporal, threshold and trends analyses of an economic case-study of European Union member states' competitiveness [2–4]. Further, ICrA has been used to discover the dependencies of different problems such as:

- universities ranking [5, 6];
- rivers pollution modelling [7, 8];
- neural networks [9];
- properties of crude oils [10];
- blood plasma analysis [11];
- Behterev's disease analysis [12].

ICrA approach has proved itself as a very useful tool in the case of modeling of fermentation processes (FP). ICrA is applied for establishing the relations between genetic algorithm (GA) parameters (generation gap, number of individuals, crossover rate, mutation rate) on the one hand, and convergence time, model accuracy and model parameters on the other hand in case of *E. coli* FP [13, 14] and *S. cerevisiae* FP [15–18]. In [19, 20] ICrA is applied to establish the dependencies of considered parameters based on different criteria referred to various metaheuristic algorithms, namely Ant Colony Optimization (ACO) hybrid schemes using GA and ACO.

In the meantime, based on the numerous applications of ICrA, the underlying theory is being continuously developed. Thus far, five algorithms for InterCriteria relations calculation – μ-biased, Balanced, ν-biased, Unbiased adn Weighted – are proposed [21]. The proposed ICrA algorithms are realized based on the ideas presented in [22]. In [22] several rules for defining the ways of estimating the degrees of "agreement" and "disagreement", with respect to the type of data, are proposed. It is shown that it is necessary to specify the algorithms for determining the degrees of "agreement" and "disagreement". However, with few exceptions, in all ICrA application the μ-biased algorithm is used. But in some case studies the μ-biased algorithm, as well as three of the the other ICrA algorithms (Balanced, ν-biased and Unbiased) may not be suitable. The main purpose of this research is to show that the Weighted algorithm is appropriate for analysis of data with numerous identical values (predominantly repetitive data) which are natural for the problem under consideration. In such cases, the other ICrA algorithms will consider these identical values as uncertainty and will not be able to correctly evaluate the relation between the criteria examined.

In this investigation all five ICrA algorithms, μ-biased, Balanced, ν-biased, Unbiased and Weighted are applied to explore the correlations between *E. coli* fed-batch FP model parameters (μ_{max}, k_S and $Y_{S/X}$), on the one side, and genetic algorithm outcomes – objective function value J and computation time T, on the other side. The obtained results for the considered ICrA criteria dependencies are compared and the most reliable algorithm for InterCriteria relations calculation has been distinguished.

The cross-platform software for ICrA approach [23], ICrAData is used to perform ICrA. The rest of the paper is organized as follows. In Sect. 2 the theoretical background of the ICrA is given, as well as the different algorithms for Intercriteria relations calculation. In Sect. 3 the investigated case study, parameter identification problem, is presented. The numerical results for identification procedures using genetic algorithms are discussed in Sect. 4. In Sect. 5 the results from application of ICrA are presented and the performance of the different InterCriteria relations calculation are discussed. Concluding remarks are presented in Sect. 6.

2 InterCriteria Analysis

2.1 Theoretical Background

InterCriteria analysis, based on the apparatuses of Index Matrices (IM) [24–28] and Intuitionistic Fuzzy Sets (IFS) [29–31], is given in details in [1]. Here, for completeness, the proposed idea is briefly presented.

Let the initial IM have the form of Eq. (1), where, for every p, q, $(1 \leq p \leq m, 1 \leq q \leq n)$, C_p is a criterion, taking part in the evaluation; O_q – an object to be evaluated; $C_p(O_q)$ – a real number (the value assigned by the p-th criteria to the q-th object).

$$
A = \begin{array}{c|ccccc}
 & O_1 & \dots & O_q & \dots & O_n \\
\hline
C_1 & C_1(O_1) & \dots & C_1(O_q) & \dots & C_1(O_n) \\
\vdots & \vdots & \ddots & \vdots & \ddots & \vdots \\
C_p & C_p(O_1) & \dots & C_p(O_q) & \dots & C_p(O_n) \\
\vdots & \vdots & \ddots & \vdots & \ddots & \vdots \\
C_m & C_m(O_1) & \dots & C_m(O_q) & \dots & C_m(O_n)
\end{array}
\tag{1}
$$

Let O denote the set of all objects being evaluated, and $C(O)$ be the set of values assigned by a given criteria C (i.e., $C = C_p$ for some fixed p) to the objects, i.e.,

$$O \stackrel{\text{def}}{=} \{O_1, O_2, O_3, \dots, O_n\},$$
$$C(O) \stackrel{\text{def}}{=} \{C(O_1), C(O_2), C(O_3), \dots, C(O_n)\}.$$

Let $x_i = C(O_i)$. Then the following set can be defined:

$$C^*(O) \stackrel{\text{def}}{=} \{\langle x_i, x_j \rangle | i \neq j \,\&\, \langle x_i, x_j \rangle \in C(O) \times C(O)\}.$$

Further, if $x = C(O_i)$ and $y = C(O_j)$, $x \prec y$ will be written iff $i < j$.

In order to find the agreement between two criteria, the vectors of all internal comparisons for each criterion are constructed, which elements fulfil one of the three relations R, $\overline{R}$ and $\tilde{R}$. The nature of the relations is chosen such that for a fixed criterion C and any ordered pair $\langle x, y \rangle \in C^*(O)$:

$$\langle x, y \rangle \in R \Leftrightarrow \langle y, x \rangle \in \overline{R}, \tag{2}$$

$$\langle x, y \rangle \in \tilde{R} \Leftrightarrow \langle x, y \rangle \notin (R \cup \overline{R}), \tag{3}$$

$$R \cup \overline{R} \cup \tilde{R} = C^*(O). \tag{4}$$

Further, we consider R, $\overline{R}$ and $\tilde{R}$ to be $>$, $<$ and $=$, respectively. For the effective calculation of the vector of internal comparisons (denoted further by $V(C)$) only the subset of $C(O) \times C(O)$ needs to be considered, namely:

$$C^{\prec}(O) \overset{\text{def}}{=} \{\langle x, y \rangle \mid x \prec y \ \& \ \langle x, y \rangle \in C(O) \times C(O),$$

due to Eqs. (2)–(4). For brevity, $c^{i,j} = \langle C(O_i), C(O_j) \rangle$.

Then for a fixed criterion C the vector of lexicographically ordered pair elements is constructed:

$$V(C) = \{c^{1,2}, c^{1,3}, \ldots, c^{1,n}, c^{2,3}, c^{2,4}, \ldots, c^{2,n}, c^{3,4}, \ldots, c^{3,n}, \ldots, c^{n-1,n}\}. \tag{5}$$

In order to be more suitable for calculations, $V(C)$ is replaced by $\hat{V}(C)$, where its k-th component ($1 \leq k \leq \frac{n(n-1)}{2}$) is given by:

$$\hat{V}_k(C) = \begin{cases} 1, & \text{iff } V_k(C) \in R, \\ -1, & \text{iff } V_k(C) \in \overline{R}, \\ 0, & \text{otherwise.} \end{cases}$$

When comparing two criteria the degree of "agreement" ($\mu_{C,C'}$) is usually determined as the number of matching components of the respective vectors. The degree of "disagreement" ($\nu_{C,C'}$) is usually the number of components of opposing signs in the two vectors. From the way of computation it is obvious that $\mu_{C,C'} = \mu_{C',C}$ and $\nu_{C,C'} = \nu_{C',C}$. Moreover, $\langle \mu_{C,C'}, \nu_{C,C'} \rangle$ is an Intuitionistic Fuzzy Pair (IFP) [32].

There may be some pairs $\langle \mu_{C,C'}, \nu_{C,C'} \rangle$, for which the sum $\mu_{C,C'} + \nu_{C,C'}$ is less than 1. The difference

$$\pi_{C,C'} = 1 - \mu_{C,C'} - \nu_{C,C'} \tag{6}$$

is considered as a degree of "uncertainty".

2.2   ICrA Algorithms for Calculation of $\mu_{C,C'}$ and $\nu_{C,C'}$

Five different algorithms for calculation of $\mu_{C,C'}$ and $\nu_{C,C'}$ are known. They are based on the ideas presented in [22], as follows:

- **μ-biased**: This algorithm follows the rules presented in [22, Table 3], where the rule $=, =$ for two criteria C and C' is assigned to $\mu_{C,C'}$. An example pseudocode is presented below as **Algorithm 1**.
- **Unbiased**: This algorithm follows the rules in [22, Table 1]. It should be noted that in such case a criterion compared to itself does not necessarily yield $\langle 1, 0 \rangle$, too. An example pseudocode is presented below as **Algorithm 2**.
- **ν-biased**: In this case the rule $=, =$ for two criteria C and C' is assigned to $\nu_{C,C'}$. It should be noted that in such case a criteria compared to itself does not necessarily yield $\langle 1, 0 \rangle$. An example pseudocode is presented below as **Algorithm 3**.
- **Balanced**: This algorithm follows the rules in [22, Table 2], where the rule $=, =$ for two criteria C and C' is assigned a half to both $\mu_{C,C'}$ and $\nu_{C,C'}$. It should be noted that in such case a criteria compared to itself does not necessarily yield $\langle 1, 0 \rangle$. An example pseudocode is presented below as **Algorithm 4**.
- **Weighted**: The algorithm is based on the Unbiased algorithm for the initial estimation of the degrees of "agreement" $\mu_{C,C'}$ and the degrees of "disagreement" $\nu_{C,C'}$, however, at the end of it the values of $\pi_{C,C'}$ are proportionally distributed to $\mu_{C,C'}$ and $\nu_{C,C'}$. Thus, the final values of $\mu_{C,C'}$ and $\nu_{C,C'}$ generated by this algorithm will always complement to 1. An example pseudocode is presented below as **Algorithm 5**.

Algorithm 1 : μ-biased

Require: Vectors $\hat{V}(C)$ and $\hat{V}(C')$

1: **function** DEGREES OF AGREEMENT AND DISAGREEMENT($\hat{V}(C)$, $\hat{V}(C')$)
2:      $V \leftarrow \hat{V}(C) - \hat{V}(C')$
3:      $\mu \leftarrow 0$
4:      $\nu \leftarrow 0$
5:      **for** $i \leftarrow 1$ to $\frac{n(n-1)}{2}$ **do**
6:        **if** $V_i = 0$ **then**
7:          $\mu \leftarrow \mu + 1$
8:        **else if** $abs(V_i) = 2$ **then**   ▷ $abs(V_i)$: the absolute value of V_i
9:          $\nu \leftarrow \nu + 1$
10:       **end if**
11:     **end for**
12:     $\mu \leftarrow \frac{2}{n(n-1)} \mu$
13:     $\nu \leftarrow \frac{2}{n(n-1)} \nu$
14:     **return** μ, ν
15: **end function**

As a result of applying any of the proposed algorithms to IM A (Eq. (1)), the following IM is obtained:

$$\begin{array}{c|ccc} & C_2 & \cdots & C_m \\ \hline C_1 & \langle \mu_{C_1,C_2}, \nu_{C_1,C_2} \rangle & \cdots & \langle \mu_{C_1,C_m}, \nu_{C_1,C_m} \rangle \\ \vdots & \vdots & \ddots & \vdots \\ C_{m-1} & & \cdots & \langle \mu_{C_{m-1},C_m}, \nu_{C_{m-1},C_m} \rangle \end{array},$$

that determines the degrees of "agreement" and "disagreement" between criteria $C_1, ..., C_m$.

Algorithm 2 : Unbiased

Require: Vectors $\hat{V}(C)$ and $\hat{V}(C')$

1: **function** DEGREES OF AGREEMENT AND DISAGREEMENT($\hat{V}(C)$, $\hat{V}(C')$)
2:     $P \leftarrow \hat{V}(C) \odot \hat{V}(C')$ ▷ $\odot$ denotes Hadamard (entrywise) product
3:     $V \leftarrow \hat{V}(C) - \hat{V}(C')$
4:     $\mu \leftarrow 0$
5:     $\nu \leftarrow 0$
6:     **for** $i \leftarrow 1$ to $\frac{n(n-1)}{2}$ **do**
7:         **if** $V_i = 0$ and $P_i \neq 0$ **then**
8:             $\mu \leftarrow \mu + 1$
9:         **else if** abs(V_i) = 2 **then**   ▷ abs(V_i): the absolute value of V_i
10:            $\nu \leftarrow \nu + 1$
11:        **end if**
12:    **end for**
13:    $\mu \leftarrow \frac{2}{n(n-1)} \mu$
14:    $\nu \leftarrow \frac{2}{n(n-1)} \nu$
15:    **return** μ, ν
16: **end function**

Algorithm 3 : ν-biased

Require: Vectors $\hat{V}(C)$ and $\hat{V}(C')$

1: **function** DEGREES OF AGREEMENT AND DISAGREEMENT($\hat{V}(C)$, $\hat{V}(C')$)
2:     $P \leftarrow \hat{V}(C) \odot \hat{V}(C')$ ▷ $\odot$ denotes Hadamard (entrywise) product
3:     $V \leftarrow \hat{V}(C) - \hat{V}(C')$
4:     $\mu \leftarrow 0$
5:     $\nu \leftarrow 0$
6:     **for** $i \leftarrow 1$ to $\frac{n(n-1)}{2}$ **do**
7:         **if** $V_i = 0$ and $P_i \neq 0$ **then**
8:             $\mu \leftarrow \mu + 1$
9:         **else if** $V_i = P_i$ or abs(V_i) = 2 **then**   ▷ abs(V_i): the absolute value of V_i
10:            $\nu \leftarrow \nu + 1$
11:        **end if**
12:    **end for**
13:    $\mu \leftarrow \frac{2}{n(n-1)} \mu$
14:    $\nu \leftarrow \frac{2}{n(n-1)} \nu$
15:    **return** μ, ν
16: **end function**

Algorithm 4 : Balanced

Require: Vectors $\hat{V}(C)$ and $\hat{V}(C')$

1: **function** DEGREES OF AGREEMENT AND DISAGREEMENT($\hat{V}(C), \hat{V}(C')$)
2: $\quad P \leftarrow \hat{V}(C) \odot \hat{V}(C')$ $\quad\quad\quad\quad\quad\quad\quad\quad\quad$ ▷ $\odot$ denotes Hadamard (entrywise) product
3: $\quad V \leftarrow \hat{V}(C) - \hat{V}(C')$
4: $\quad \mu \leftarrow 0$
5: $\quad \nu \leftarrow 0$
6: $\quad$ **for** $i \leftarrow 1$ to $\frac{n(n-1)}{2}$ **do**
7: $\quad\quad$ **if** $V_i = P_i$ **then**
8: $\quad\quad\quad \mu \leftarrow \mu + \frac{1}{2}$
9: $\quad\quad\quad \nu \leftarrow \nu + \frac{1}{2}$
10: $\quad\quad$ **else if** $V_i = 0$ and $P_i \neq 0$ **then**
11: $\quad\quad\quad \mu \leftarrow \mu + 1$
12: $\quad\quad$ **else if** abs(V_i) = 2 **then** $\quad\quad\quad\quad\quad\quad$ ▷ abs(V_i): the absolute value of V_i
13: $\quad\quad\quad \nu \leftarrow \nu + 1$
14: $\quad\quad$ **end if**
15: $\quad$ **end for**
16: $\quad \mu \leftarrow \frac{2}{n(n-1)} \mu$
17: $\quad \nu \leftarrow \frac{2}{n(n-1)} \nu$
18: $\quad$ **return** μ, ν
19: **end function**

3 Case Study: Parameter Identification of an *E. coli* Fed-batch Fermentation Process Model

3.1 Mathematical Model

Let us use the following non-linear differential equation system to describe the *E. coli* fed-batch FP [20]:

$$\frac{dX}{dt} = \mu X - \frac{F_{in}}{V} X, \tag{7}$$

$$\frac{dS}{dt} = -q_S X + \frac{F_{in}}{V} (S_{in} - S), \tag{8}$$

$$\frac{dV}{dt} = F_{in}, \tag{9}$$

where

$$\mu = \mu_{max} \frac{S}{k_S + S}, \tag{10}$$

$$q_S = \frac{1}{Y_{S/X}} \mu \tag{11}$$

Algorithm 5 : Weighted

Require: Vectors $\hat{V}(C)$ and $\hat{V}(C')$

1: **function** DEGREES OF AGREEMENT AND DISAGREEMENT($\hat{V}(C)$, $\hat{V}(C')$)
2:     $P \leftarrow \hat{V}(C) \odot \hat{V}(C')$ ▷ $\odot$ denotes Hadamard (entrywise) product
3:     $V \leftarrow \hat{V}(C) - \hat{V}(C')$
4:     $\mu \leftarrow 0$
5:     $\nu \leftarrow 0$
6:     **for** $i \leftarrow 1$ to $\frac{n(n-1)}{2}$ **do**
7:         **if** $V_i = 0$ and $P_i \neq 0$ **then**
8:             $\mu \leftarrow \mu + 1$
9:         **else if** abs(V_i) = 2 **then**   ▷ abs(V_i): the absolute value of V_i
10:             $\nu \leftarrow \nu + 1$
11:         **end if**
12:     **end for**
13:     $\mu \leftarrow \frac{2}{n(n-1)} \mu$
14:     $\nu \leftarrow \frac{2}{n(n-1)} \nu$
15:     **if** $\mu + \nu \neq 0$ **then**
16:         $\mu \leftarrow \frac{\mu}{\mu+\nu}$
17:         $\nu \leftarrow \frac{\mu}{\mu+\nu}$
18:     **else**
19:         $\mu \leftarrow \frac{1}{2}$
20:         $\nu \leftarrow \frac{1}{2}$
21:     **end if**
22:     **return** μ, ν
23: **end function**

---

and X is the biomass concentration, [g/l];
S – the substrate concentration, [g/l];
F_{in} – the feeding rate, [l/h];
V – the bioreactor volume, [l];
S_{in} – the substrate concentration in the feeding solution, [g/l];
μ and q_S – the specific rate functions, [1/h];
μ_{max} – the maximum value of the μ, [1/h];
k_S – the saturation constant, [g/l];
$Y_{S/X}$ – the yield coefficient, [g/g].

For the model (Eqs. 7–11) the parameters that will be identified are μ_{max}, k_S and $Y_{S/X}$.

For the model parameters identification we use experimental data for biomass and glucose concentration of an *E. coli* MC4110 fed-batch fermentation process. The detailed description of the process condition and experimental data are presented in [20].

3.2 Genetic Algorithms

Genetic algorithms (GAs) are used to identify the parameters in the *E. coli* fed-batch FP model (Eqs. 7–11). GA was developed to model adaptation processes mainly operating on binary strings and using a recombination operator with mutation as a background operator. The GA maintains a population of chromosomes, $P(t) = x_1^t, ..., x_n^t$ for generation t. Each chromosome represents a potential solution to the problem and is implemented as some data structure S. Each solution is evaluated to give some measure of its "fitness". Fitness of an chromosome is assigned proportionally to the value of the objective function of the chromosomes. Then, a new population (generation $t + 1$) is formed by selecting more fit chromosomes (selection step). Some members of the new population undergo transformations by means of "genetic" operators to form new solution. There are unary transformations m_i (mutation type), which create new chromosomes by a small change in a single chromosome ($m_i : S \rightarrow S$), and higher order transformations c_j (crossover type), which create new chromosomes by combining parts from several chromosomes ($c_j : S \times ... \times S \rightarrow S$). After some number of generations the algorithm converges – it is expected that the best chromosome represents a near-optimum (reasonable) solution. The combined effect of selection, crossover and mutation gives so-called reproductive scheme growth equation [33]:

$$\xi(S, t+1) \geq \xi(S, t) \cdot eval(S, t) / \bar{F}(t) \left[1 - p_c \cdot \frac{\delta(S)}{m-1} - o(S) \cdot p_m \right]$$

The structure of the herewith used GA is shown by the pseudocode below (Algorithm 6).

Algorithm 6 Pseudocode for GA

1: **begin**
2:     $i \leftarrow 0$
3:     Initial population $P(0)$
4:     Evaluate $P(0)$
5:     **while** not done **do**
6:         (test for termination criterion)
7:     **end while**
8:     **begin**
9:         $i \leftarrow i + 1$
10:        Select $P(i)$ from $P(i - 1)$
11:        Recombine $P(i)$
12:        Mutate $P(i)$
13:        Evaluate $P(i)$
14:    **end**
15: **end**

The three model parameters μ_{max}, k_S and $Y_{S/X}$ are represented in the chromosome. The following upper and lower bounds of the model parameters are considered:

$$0 < \mu_{max} < 0.7,$$

$$0 < k_S < 1,$$

$$0 < Y_{S/X} < 30.$$

The probability, P_i, for each chromosome is defined by:

$$P[\text{Individual } i \text{ is chosen}] = \frac{F_i}{\sum\limits_{j=1}^{PopSize} F_j}, \tag{12}$$

where F_i equals the fitness of chromosome i and $PopSize$ is the population size.

For the considered here model parameter identification, the type of the basic operators in GA are as follows:

- encoding – binary,
- fitness function – linear ranking,
- selection function – roulette wheel selection,
- crossover function – two point crossover,
- mutation function – binary mutation,
- reinsertion – fitness-based.

The values of GA parameters are:

- generation gap, ggap = 0.97,
- crossover probability, xovr = 0.75,
- mutation probability, mutr = 0.01,
- maximum number of generations, maxgen = 200.

The optimization criterion is a certain factor, whose value defines the quality of an estimated set of parameters.

The objective consists of adjusting the parameters (μ_{max}, k_S and $Y_{S/X}$) of the non-linear mathematical model function (Eqs. 7–11) to best fit a data set. To evaluate the mismatch between experimental and model predicted data the Least Square Regression is used:

$$J = \sum (S_{exp} - S_{mod})^2 + (X_{exp} - X_{mod})^2 \rightarrow min, \tag{13}$$

where X_{exp} and S_{exp} are the real experimental data; X_{mod} and S_{mod} are the model predicted data.

4 Numerical Results of Model Parameter Identification Using Genetic Algorithms

All computations are performed using a PC/Intel Core i5-2320 CPU @ 3.00GHz, 8 GB Memory (RAM), Windows 8.1 (64 bit) operating system and Matlab 7.5 environment.

Fourteen differently tuned GA are applied consistently to estimate the model parameters. In each GA various population sizes – from 5 to 200 chromosomes in the population, are used as follows: GA with 5, 10, 20, 30, ..., 100, 110, 150 and 200 chromosomes in the population. For all 14 GAs the values of the main algorithm parameters are as presented in Sect. 3.2.

Using mathematical model of the *E. coli* cultivation process (Eqs. 7–11) the model parameters – maximum specific growth rate (μ_{max}), saturation constant (k_S) and yield coefficient ($Y_{S/X}$) – are estimated based on series of numerical experiments. For the identification procedures consistently different population sizes (from 5 to 200 chromosomes in the population) are used. The number of generation is fixed to 200. Because of the stochastic characteristics of the applied GA series of 30 runs for each population size are performed. In Table 1, the obtained average results for the model parameters (μ_{max}, k_S and $Y_{S/X}$), objective function value J (Eq. 13) and computational time T are presented.

The results show that for the model parameter μ_{max} the GAs obtain five different values – 0.55, 0.53, 0.51, 0.50 and 0.49. The last two values are obtained by five different GAs each. For parameter k_S two values are observed – 0.01 (nine results) and 0.02 (five results). The obtained values for model parameter $Y_{X/S}$ are two – 2.03

Table 1 Average results from the model parameter identification

	μ_{max} (g/l)	k_S (g/l)	$Y_{S/X}$ (g/g)	J	T (s)
GA_1	0.55	0.02	2.03	6.27	4.65
GA_2	0.53	0.02	2.03	5.84	6.05
GA_3	0.51	0.02	2.02	4.76	7.47
GA_4	0.49	0.01	2.02	4.56	11.25
GA_5	0.49	0.01	2.02	4.65	12.92
GA_6	0.51	0.02	2.02	4.61	14.65
GA_7	0.50	0.01	2.02	4.58	16.97
GA_8	0.50	0.01	2.02	4.57	19.72
GA_9	0.50	0.01	2.02	4.58	21.79
GA_{10}	0.50	0.01	2.02	4.57	24.20
GA_{11}	0.49	0.01	2.02	4.55	26.85
GA_{12}	0.50	0.02	2.02	4.55	29.52
GA_{13}	0.49	0.01	2.02	4.56	39.41
GA_{14}	0.49	0.01	2.02	4.55	51.92

(two results) and 2.02 (twelve results). Based on these results it could be concluded that most of the considered GAs show similar performance. Considering the objective function values it is clear that GAs from GA_7 to GA_{14} perform better than the first six GAs (see Table 1). In order to obtain a model with high degree of accuracy the used population from individuals should consist at least of 60 chromosomes. Based on the observed J and T values, for the considered here problem the optimum population size is 100.

In Figs. 1 and 2 the comparison of the model predictions (obtained by model parameters estimated by GA_{11}) and real experimental data of *E. coli* fed-batch fermentation process.

The presented Figs. 1 and 2 illustrate a very good correlation between the experimental and predicted data of the considered process variables. The obtained results

Fig. 1 Biomass dynamics – model versus experimental data

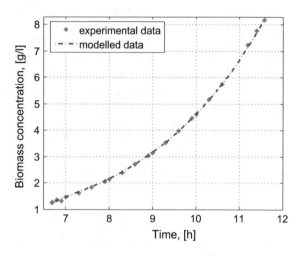

Fig. 2 Glucose dynamics – model versus experimental data

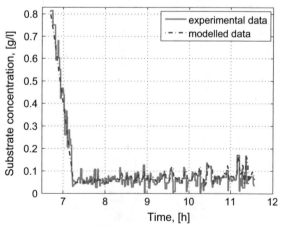

show the GA_{11} algorithm performs very well for model parameter identification of an *E. coli* fed-batch fermentation process.

5 ICrA of the Numerical Results from Parameter Identification Problem

In order to establish the relations and dependencies between the parameters in a non-linear model of an *E. coli* fed-batch cultivation process and objective function J values and computation times T, ICrA is performed.

A new cross-platform software for ICrA approach, ICrAData is used in order to perform ICrA. In the ICrAData software [23], available at [34], all five different algorithms for InterCriteria relations calculation, namely μ-biased, Unbiased, ν-biased, Balanced and Weighted, are implemented.

Based on the obtained data from the all model parameters identification procedures (Table 1), an IM is constructed. The ICrA criteria are defined, as follows: criteria $C1$, $C2$ and $C3$ are the values of *E. coli* model parameters estimates (μ_{max}, k_S and $Y_{S/X}$, respectively); $C4$ is the objective function value (J) and $C5$ is the computation time (T). Fourteen differently tuned GAs are taken as ICrA objects, as follows: $GA_1, GA_2, GA_3, GA_4, \ldots, GA_{11}, GA_{12}, GA_{13}, GA_{14}$, respectively GA with 5, 10, 20, 30,..., 100, 110, 150 and 200 chromosomes. The constructed IM has the form:

$$
\begin{array}{c|ccccc}
 & GA_1 & GA_2 & \ldots & GA_{13} & GA_{14} \\
\hline
C_1 & 0.55 & 0.53 & \ldots & 0.49 & 0.49 \\
\vdots & \vdots & \vdots & \vdots & \vdots & \vdots \\
C_5 & 4.65 & 6.05 & \ldots & 39.41 & 51.92
\end{array}
\tag{14}
$$

This example is chosen as particularly appropriate to test the performance of the five different ICrA algorithms (μ-biased, Unbiased, ν-biased, Balanced and the Weighted algorithm). As it can be seen from the results presented in Table 1, there are many identical results for the model parameter estimates. These results are legitimate outcomes of GA optimization. It is desirable that GA obtains the same estimates for a given model parameter. So in this case, these identical values are not signifying uncertainty. Since the five ICrA algorithms differ exactly in the way they treat identical results, this example is particularly indicative.

All five ICrA algorithms are applied using constructed IM (Eq. 14). The obtained results for the degree of "agreement" $\mu_{C,C'}$ and the degree of "disagreement" $\nu_{C,C'}$ are summarized in Table 2.

The Euclidean distances from each pair to point $\langle 1, 0 \rangle$, calculated by ICrAData, as proposed in [35], are given in Table 3.

Table 2 Results from ICrA – pairs $\langle \mu_{C,C'}, \nu_{C,C'} \rangle$

Criteria pair	μ-biased	Unbiased	ν-biased	Balanced	Weighted
$C1 \leftrightarrow C2$	$\langle 0.637, 0.000 \rangle$	$\langle 0.451, 0.000 \rangle$	$\langle 0.451, 0.187 \rangle$	$\langle 0.544, 0.093 \rangle$	$\langle 1.000, 0.000 \rangle$
$C1 \leftrightarrow C3$	$\langle 0.495, 0.000 \rangle$	$\langle 0.264, 0.000 \rangle$	$\langle 0.264, 0.231 \rangle$	$\langle 0.379, 0.115 \rangle$	$\langle 1.000, 0.000 \rangle$
$C1 \leftrightarrow C4$	$\langle 0.703, 0.088 \rangle$	$\langle 0.660, 0.088 \rangle$	$\langle 0.659, 0.132 \rangle$	$\langle 0.681, 0.110 \rangle$	$\langle 0.882, 0.118 \rangle$
$C1 \leftrightarrow C5$	$\langle 0.143, 0.626 \rangle$	$\langle 0.143, 0.626 \rangle$	$\langle 0.143, 0.626 \rangle$	$\langle 0.143, 0.626 \rangle$	$\langle 0.186, 0.814 \rangle$
$C2 \leftrightarrow C3$	$\langle 0.637, 0.000 \rangle$	$\langle 0.198, 0.000 \rangle$	$\langle 0.198, 0.440 \rangle$	$\langle 0.418, 0.220 \rangle$	$\langle 1.000, 0.000 \rangle$
$C2 \leftrightarrow C4$	$\langle 0.429, 0.088 \rangle$	$\langle 0.385, 0.088 \rangle$	$\langle 0.385, 0.132 \rangle$	$\langle 0.407, 0.110 \rangle$	$\langle 0.814, 0.186 \rangle$
$C2 \leftrightarrow C5$	$\langle 0.099, 0.396 \rangle$	$\langle 0.099, 0.396 \rangle$	$\langle 0.099, 0.396 \rangle$	$\langle 0.099, 0.396 \rangle$	$\langle 0.200, 0.800 \rangle$
$C3 \leftrightarrow C4$	$\langle 0.330, 0.000 \rangle$	$\langle 0.264, 0.000 \rangle$	$\langle 0.264, 0.066 \rangle$	$\langle 0.297, 0.033 \rangle$	$\langle 1.000, 0.000 \rangle$
$C3 \leftrightarrow C5$	$\langle 0.000, 0.264 \rangle$	$\langle 0.000, 0.264 \rangle$	$\langle 0.000, 0.264 \rangle$	$\langle 0.000, 0.264 \rangle$	$\langle 0.000, 1.000 \rangle$
$C4 \leftrightarrow C5$	$\langle 0.099, 0.835 \rangle$	$\langle 0.099, 0.835 \rangle$	$\langle 0.099, 0.835 \rangle$	$\langle 0.099, 0.835 \rangle$	$\langle 0.106, 0.894 \rangle$

Table 3 Results from ICrA Euclidean distance from each pair to $\langle 1, 0 \rangle$ point

Criteria pair	μ-biased	Unbiased	ν-biased	Balanced	Weighted
$C1 \leftrightarrow C2$	0.3626	0.5495	0.5803	0.4655	0.0000
$C1 \leftrightarrow C3$	0.5055	0.7363	0.7716	0.6315	0.0000
$C1 \leftrightarrow C4$	0.3095	0.3518	0.3653	0.3371	0.1664
$C1 \leftrightarrow C5$	1.0616	1.0616	1.0616	1.0616	1.1516
$C2 \leftrightarrow C3$	0.3626	0.8022	0.9147	0.6225	0.0000
$C2 \leftrightarrow C4$	0.5782	0.6216	0.6294	0.6035	0.2631
$C2 \leftrightarrow C5$	0.9841	0.9841	0.9841	0.9841	1.1314
$C3 \leftrightarrow C4$	0.6703	0.7363	0.7392	0.7041	0.0000
$C3 \leftrightarrow C5$	1.0342	1.0342	1.0342	1.0342	1.4142
$C4 \leftrightarrow C5$	1.2286	1.2286	1.2286	1.2286	1.2645

The next three figures (Figs. 3, 4 and 5) present screenshots of the ICrAData software. In Fig. 3 ICrA results by ICrAData, applying μ-biased algorithm, in primary view (matrices with $\mu_{C,C'}$ and $\nu_{C,C'}$ values and graphic interpretation within the intuitionistic fuzzy triangle (triangle view) are presented. In Fig. 4 ICrA results by ICrAData, applying Weighted algorithm, in primary view (matrices with $\mu_{C,C'}$ and $\nu_{C,C'}$ values) and graphic interpretation within the intuitionistic fuzzy triangle (text view) are presented. For the same algorithm, results of secondary view (matrix with $\langle \mu_{C,C'}, \nu_{C,C'} \rangle$ values, Euclidean distance of each criteria pair from point $\langle 1, 0 \rangle$) and graphic interpretation within the intuitionistic fuzzy triangle (text view) are shown in Fig. 5.

The results obtained from the five ICrA algorithms, visualized in the intuitionistic fuzzy interpretation triangle (IFIT), are presented in Figs. 6, 7, 8, 9 and 10.

As it can be seen there are some differences in the obtained results mainly with respect to the criteria with identical evaluation for several objects. In this example,

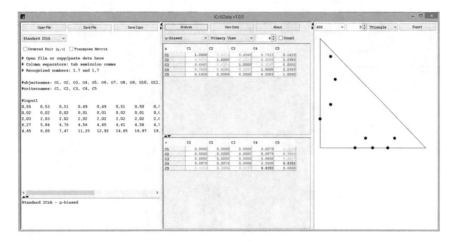

Fig. 3 Primary view of the results based on μ-biased algorithm

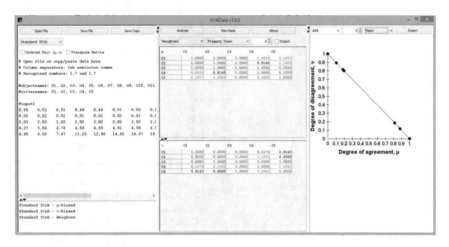

Fig. 4 Primary view of the results based on Weighted algorithm

such results are found in the estimate of the three model parameters (μ_{max} ($C1$), k_S ($C2$) and $Y_{S/X}$ ($C3$)) and the objective function values, J ($C4$), so the applied ICrA algorithms exhibit different behaviour for the pairs that include these criteria.

The criteria pair that exhibits distinctly different behavior is $C2 \leftrightarrow C3$. The calculated Euclidean distances from pair $C2 \leftrightarrow C3$ to point $\langle 1, 0 \rangle$ are between 0 and 0.91. The explanation is the greater number of identical results in the estimates of the model parameters k_S and $Y_{S/X}$. According to the scale, presented in [22], in the case of μ-biased and Balanced ICrA algorithms this pair is in dissonance, i.e., these two criteria are not connected or correlated. In the case of Unbiased and v-biased algorithms, the pair is in weak negative consonance, i.e., the criteria $C2$ and $C3$ are connected in some kind of inverse dependence. The Weighted algorithm shows that

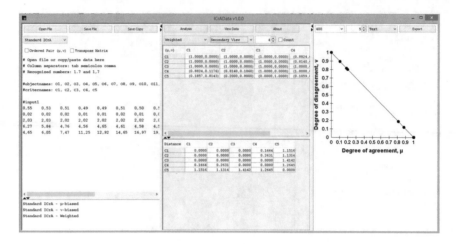

Fig. 5 Secondary view of the results based on Weighted algorithm

Fig. 6 Presentation of ICrA
results by pairs in the IFIT –
μ-biased algorithm

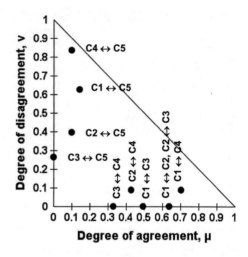

the pair is in strong (strongest) positive consonance, i.e., these two criteria are in
great dependence or are strongly correlated. So, which is true? If we look from the
perspective of physical meaning of these model parameters [36], they should be in
positive consonance. Since the rate of substrate utilization is related to the specific
growth rate, it is obvious that the yield coefficient $Y_{S/X}$ (mass of cells produced per
unit mass of substrate utilized) is correlated to the saturation constant k_S from the
specific growth rate equation [37]. This fact shows that in the case of the presence
of the identical (legitimate) data only the Weighted ICrA algorithm gives the correct
evaluation of the pair dependence. The rest of the ICrA algorithm accept the identical
data as inherent uncertainty and produce contradictory results.

Let us consider the following criteria pairs $C1 \leftrightarrow C5$, $C2 \leftrightarrow C5$, $C3 \leftrightarrow C5$ and
$C4 \leftrightarrow C5$, i.e., relation between computation time T ($C5$) and the remaining param-

Fig. 7 Presentation of ICrA results by pairs in the IFIT – Unbiased algorithm

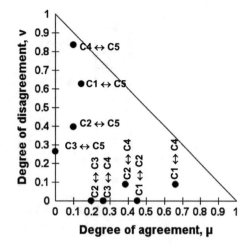

Fig. 8 Presentation of ICrA results by pairs in the IFIT – v-biased algorithm

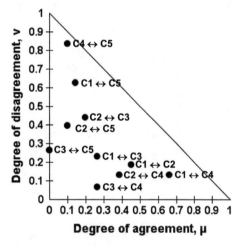

eters μ_{max} ($C1$), k_S ($C2$), $Y_{S/X}$ ($C3$) and J ($C4$)). Since there are no identical data in the values of the computation time, all five ICrA algorithms give similar results (see Tables 2 and 3). The obtained value of degree of "agreement" and the degree of "disagreement" are the same for the μ-biased, Unbiased, v-biased and Balanced ICrA algorithms. The Weighted algorithm produces similar results for the degree of "agreement". Due to the elimination of the degree of "uncertainty" the results for the degree of "disagreement" are different. There is only one exception – the results for the pair $C4 \leftrightarrow C5$, because of the lesser number of identical data in the $C4$ criteria values. Generally, the observed differences are insignificant and all the pairs $C1 \leftrightarrow C5$, $C2 \leftrightarrow C5$ and $C4 \leftrightarrow C5$ are in negative consonance, and the pair $C3 \leftrightarrow C5$ is even in strong negative consonance. Altogether, such results are logical – a large number of GA algorithm iterations are leading to greater computation time

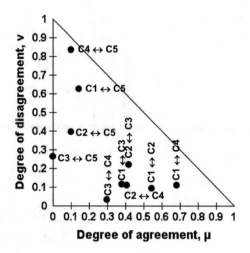

Fig. 9 Presentation of ICrA results by pairs in the IFIT – Balanced algorithm

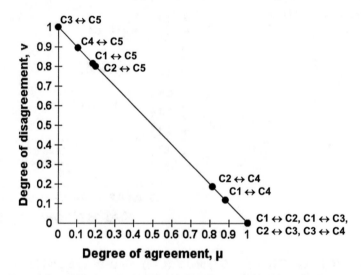

Fig. 10 Presentation of ICrA results by pairs in the IFIT – Weighted algorithm

and a more accurate solution which is reflected by smaller value of J. Likewise, the better the objective function value the better the model parameter estimates. In summary, for these criteria pairs all ICrA algorithms perform correctly.

Considering the criteria pairs $C1 \leftrightarrow C2$ and $C1 \leftrightarrow C3$, a very strong correlation (positive consonance) is expected due to the physical meaning of these criteria. Model parameters μ_{max} and k_S are highly dependent because they are empirical coefficients to the Monod equation [37]. Model parameters μ_{max} and $Y_{S/X}$ are also highly dependent, since the rate of substrate utilization is related to the specific growth rate, as has been already discussed above. Based on the previous analysis, it is natural that only the Weighted ICrA algorithm observes strong positive consonance. Again, due to numerous identical data for these two model parameters estimates according to the μ-biased, Unbiased, ν-biased and Balanced algorithms the pair $C1 \leftrightarrow C2$ is in dissonance.

Analyzing the correlation of the rest three criteria pairs, $C1 \leftrightarrow C4$, $C2 \leftrightarrow C4$ and $C3 \leftrightarrow C4$ we should take into account that the value of the objective function J ($C4$) depends precisely on the estimates of the three model parameters (μ_{max}, k_S and $Y_{S/X}$) [36, 38]. So, the positive consonance for the criteria pairs is anticipated. As it should be expected the Weighted ICrA algorithm produces the correct estimates. All the pairs are in consonance: $C1 \leftrightarrow C4$ is in positive consonance, $C2 \leftrightarrow C4$ – weak positive consonance and $C3 \leftrightarrow C4$ strong positive consonance. The results mean that the model parameter $Y_{S/X}$ ($C3$) has greater influence on the obtained criteria value J ($C4$), i.e., the achieved model accuracy, than to the model parameters μ_{max} ($C1$) and k_S ($C2$). The sensitivity of the model parameters is important and the fact that ICrA shows that the model parameter $Y_{S/X}$ is the most sensitive, followed by μ_{max} and k_S, should be taken into account in further parameter identification procedures.

In summary, if we analyze data containing identical estimates and such estimates are expected (legitimate) the Weighted ICrA algorithm should be applied. If the identical data are due to some existing uncertainty one of μ-biased, Unbiased, ν-biased or Balanced algorithms should be used instead, in accordance with the prior knowledge of the problem.

6 Conclusion

In this paper, based on the cross-platform software ICrAData, InterCriteria analysis approach is applied. An investigation of the GA performance in a parameter identification problem of an *E. coli* fed-batch fermentation process model is considered. Based on five different algorithms for InterCriteria relations calculation, ICrA is fulfilled to evaluate the relation between the model parameters, objective function value and computation time.

Some useful recommendations regarding the selection of the most appropriate ICrA algorithm for a given data are provided. The Weighted algorithm is appropriate in the case of ICrA of data that include many identical values of criteria estimates by objects. When such results are expected due to the nature of the problem, they should not accept as uncertainty in the data. The rest of ICrA algorithms (μ-biased, Unbiased, ν-biased and Balanced) will consider these identical results as uncertainty, which will reflect to results of analysis. Thus, it is possible to obtain incorrectly identified criteria relation. Hence, it is of paramount importance to apply the most appropriate of the ICrA algorithms.

In addition, the knowledge of the relations between model parameters and GA outcomes discovered by ICrA, is valuable for future improvement of the parameter identification procedures with GAs.

Acknowledgements Work presented here is partially supported by the National Scientific Fund of Bulgaria under grants DFNI-DN 02/10 "New Instruments for Knowledge Discovery from Data, and their Modelling".

References

1. Atanassov, K., Mavrov, D., Atanassova, V.: Intercriteria decision making: a new approach for multicriteria decision making, based on index matrices and intuitionistic fuzzy sets. Issues Intuitionistic Fuzzy Sets Generalized Nets **11**, 1–8 (2014)
2. Atanassova, V., Mavrov, D., Doukovska, L., Atanassov, K.: Discussion on the threshold values in the InterCriteria decision making approach. Notes on Intuitionistic Fuzzy Sets **20**(2), 94–99 (2014)
3. Atanassova, V., Doukovska, L., Atanassov, K., Mavrov, D.: Intercriteria decision making approach to EU member states competitiveness analysis. In: Proceedings of the International Symposium on Business Modeling and Software Design - BMSD'14, pp. 289–294 (2014)
4. Atanassova, V., Doukovska, L., Karastoyanov, D., Capkovic, F.: InterCriteria decision making approach to EU member states competitiveness analysis: trend analysis. In: Intelligent Systems'2014, Advances in Intelligent Systems and Computing, vol. 322, pp. 107–115 (2014)
5. Bureva, V., Sotirova, E., Sotirov, S., Mavrov, D.: Application of the InterCriteria decision making method to Bulgarian universities ranking. Notes on Intuitionistic Fuzzy Sets **21**(2), 111–117 (2015)
6. Krawczak, M., Bureva, V., Sotirova, E., Szmidt, E.: Application of the InterCriteria decision making method to universities ranking. Adv. Intell. Syst. Comput. **401**, 365–372 (2016)
7. Ilkova, T., Petrov, M.: InterCriteria analysis for evaluation of the pollution of the Struma river in the Bulgarian section. Notes on Intuitionistic Fuzzy Sets **22**(3), 120–130 (2016)
8. Ilkova, T., Petrov, M.: Application of intercriteria analysis to the Mesta river pollution modelling. Notes on Intuitionistic Fuzzy Sets **21**(2), 118–125 (2015)
9. Sotirov, S., Sotirova, E., Melin, P., Castillo , O., Atanassov, K.: Modular neural network preprocessing procedure with intuitionistic fuzzy InterCriteria analysis method. In: Flexible Query Answering Systems 2015, Springer International Publishing, pp. 175–186 (2016)
10. Stratiev, D., Sotirov, S., Shishkova, I., Nedelchev, A., Sharafutdinov, I., Veli, A., Mitkova, M., Yordanov, D., Sotirova, E., Atanassova, V., Atanassov, K., Stratiev, D., Rudnev, N., Ribagin, S.: Investigation of relationships between bulk properties and fraction properties of crude oils by application of the Intercriteria Analysis. Pet. Sci. Technol. **34**(13), 1113–1120 (2016)
11. Todinova, S., Mavrov, D., Krumova, S., Marinov, P., Atanassova, V., Atanassov, K., Taneva, S.G.: Blood plasma thermograms dataset analysis by means of InterCriteria and correlation analyses for the case of colorectal cancer. Int. J. Bioautomation **20**(1), 115–124 (2016)
12. Zaharieva, B., Doukovska, L., Ribagin, S., Radeva, I.: InterCriteria decision making approach for behterev's disease analysis. Int. J. Bioautomation **22**(2), in press (2018)
13. Roeva, O., Fidanova, S., Vassilev, P., Gepner, P.: InterCriteria analysis of a model parameters identification using genetic algorithm. In: Proceedings of the Federated Conference on Computer Science and Information Systems **5**, 501–506 (2015)
14. Roeva, O., Vassilev, P.: InterCriteria analysis of generation gap influence on genetic algorithms performance. Adv. Intell. Syst. Comput. **401**, 301–313 (2016)
15. Angelova M.: Modified genetic algorithms and intuitionistic fuzzy logic for parameter identification of fed-batch cultivation model. Ph.D. thesis, Sofia (2014) (in Bulgarian)
16. Angelova, M., Roeva, O., Pencheva, T.: InterCriteria analysis of crossover and mutation rates relations in simple genetic algorithm. In: Proceedings of the 2015 Federated Conference on Computer Science and Information Systems, vol. 5, pp. 419–424 (2015)
17. Pencheva, T., Angelova, M., Atanassova, V., Roeva, O.: InterCriteria analysis of genetic algorithm parameters in parameter identification. Notes on Intuitionistic Fuzzy Sets **21**(2), 99–110 (2015)

18. Pencheva, T., Angelova, M., Vassilev, P., Roeva, O.: InterCriteria analysis approach to parameter identification of a fermentation process model. Adv. Intell. Syst. Comput. **401**, 385–397 (2016)
19. Fidanova, S., Roeva, O., Paprzycki, M., Gepner, P.: InterCriteria analysis of ACO start strategies. In: Proceedings of the 2016 Federated Conference on Computer Science and Information Systems, vol. 8, pp. 547–550 (2016)
20. Roeva, O., Fidanova, S., Paprzycki, M.: InterCriteria analysis of ACO and GA hybrid algorithms. Stud. Comput. Intell. **610**, 107–126 (2016)
21. Roeva, O., Vassilev, P., Angelova, M., Su, J., Pencheva, T.: Comparison of different algorithms for InterCriteria relations calculation. In: 2016 IEEE 8th International Conference on Intelligent Systems, pp. 567–572 (2016)
22. Atanassov, K., Atanassova, V., Gluhchev, G.: InterCriteria analysis: ideas and problems. Notes on Intuitionistic Fuzzy Sets **21**(1), 81–88 (2015)
23. Ikonomov, N., Vassilev, P., Roeva, O.: ICrAData software for InterCriteria analysis. Int. J. Bioautomation **22**(1), 1–10 (2018)
24. Atanassov, K.: Index Matrices: Towards an Augmented Matrix Calculus. Springer International Publishing Switzerland (2014)
25. Atanassov, K.: Generalized index matrices. C. R. de l'Academie Bulgare des Sci. **40**(11), 15–18 (1987)
26. Atanassov, K.: On index matrices, Part 1: standard cases. Adv. Stud. Contemporary Mathe. **20**(2), 291–302 (2010)
27. Atanassov, K.: On index matrices, Part 2: intuitionistic fuzzy case. Proc. Jangjeon Math. Soc. **13**(2), 121–126 (2010)
28. Atanassov, K.: On index matrices. Part 5: 3-dimensional index matrices. Advanced studies. Contemporary Math. **24**(4), 423–432 (2014)
29. Atanassov, K.: Intuitionistic fuzzy sets. VII ITKR session, Sofia, 20–23 June 1983. Reprinted: Int. J. Bioautomation **20**(S1), S1–S6 (2016)
30. Atanassov, K.: On Intuitionistic Fuzzy Sets Theory. Springer, Berlin (2012)
31. Atanassov, K.: Review and new results on intuitionistic fuzzy sets, mathematical foundations of artificial intelligence seminar, Sofia, 1988, preprint IM-MFAIS-1-88. Reprinted:: Int. J. Bioautomation **20**(S1), S7–S16 (2016)
32. Atanassov, K., Szmidt, E., Kacprzyk, J.: On intuitionistic fuzzy pairs. Notes on Intuitionistic Fuzzy Sets **19**(3), 1–13 (2013)
33. Goldberg, D.E.: Genetic algorithms in search. Optimization and machine learning. Addison Wesley Longman, London (2006)
34. InterCriteria.net, ICrAData software http://intercriteria.net/software/
35. Atanassova, V.: Interpretation in the intuitionistic fuzzy triangle of the results, obtained by the InterCriteria analysis. In: Proceedings of the 9th Conference of the European Society for Fuzzy Logic and Technology (EUSFLAT), pp. 1369–1374 (2015)
36. Bastin, G., Dochain, D.: On-line Estimation and Adaptive Control of Bioreactors. Elsevier Scientific Publications, Amsterdam (1991)
37. Monod, J.: The growth of bacterial cultures. Ann. Rev. Microbiol. **3**, 371 (1949). https://doi.org/10.1146/annurev.mi.03.100149.002103
38. Roeva, O., Vassilev, P., Fidanova, S., Paprzycki, M.: InterCriteria analysis of genetic algorithms performance. Stud. Comput. Intell. **655**, 235–260 (2016)

Author Index

© Springer Nature Switzerland AG 2019
S. Fidanova (ed.), *Recent Advances in Computational Optimization*,
Studies in Computational Intelligence 795,
https://doi.org/10.1007/978-3-319-99648-6

Printed in the United States
By Bookmasters